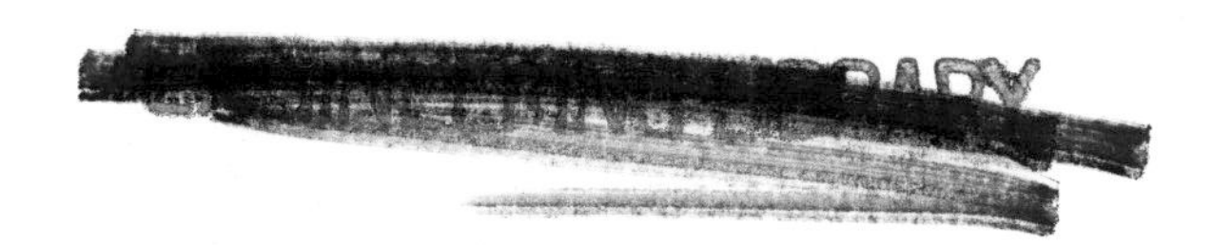

D0855969

**Essentials of UMTS**

The third generation (3G) cellular system UMTS is advanced, optimised and complex. The many existing books on UMTS attempt to explain all the intricacies of the system, and as a result are large and equally complex. This book takes a different approach and explains UMTS in a concise, clear and readily understandable style.

Written by a professional technical trainer, and based on training courses delivered on UMTS to telecommunication companies worldwide, *Essentials of UMTS* will enable you to grasp the key concepts quickly. It assumes no previous knowledge of mobile telecommunication theory, and is structured around the operation of the system, clearly setting out how the different components interact with each other, and how the system as a whole behaves. Engineers, project managers and marketing executives working for equipment manufacturers and network operators will find this concise guide to UMTS invaluable.

CHRISTOPHER COX is a technical consultant and trainer in mobile telecommunications for his business Chris Cox Communications Limited. He has a degree in Physics and a Ph.D. in Radio Astronomy from the University of Cambridge, and 15 years' experience in scientific and technical consultancy, telecommunications and training.

**The Cambridge Wireless Essentials Series**

Series Editors
WILLIAM WEBB, *Ofcom, UK*
SUDHIR DIXIT

A series of concise, practical guides for wireless industry professionals.

Martin Cave, Chris Doyle and William Webb, *Essentials of Modern Spectrum Management*
Christopher Haslett, *Essentials of Radio Wave Propagation*
Stephen Wood and Roberto Aiello, *Essentials of UWB*
Christopher Cox, *Essentials of UMTS*

*Forthcoming*
Steve Methley, *Essentials of Wireless Mesh Networking*
Linda Doyle, *Essentials of Cognitive Radio*
David Crawford, *Essentials of Mobile Television*
Malcolm Macleod and Ian Proudler, *Essentials of Smart Antennas and MIMO*
Albert Guillén í Fàbregas, *Essentials of Error Correction for Wireless Communications*

For further information on any of these titles, the series itself and ordering information see www.cambridge.org/wirelessessentials

# Essentials of UMTS

Christopher Cox
*Chris Cox Communications*

CAMBRIDGE UNIVERSITY PRESS
Cambridge, New York, Melbourne, Madrid, Cape Town, Singapore, São Paulo, Delhi

Cambridge University Press
The Edinburgh Building, Cambridge CB2 8RU, UK

Published in the United States of America by Cambridge University Press, New York

www.cambridge.org
Information on this title: www.cambridge.org/9780521889315



First published 2008

Printed in the United Kingdom at the University Press, Cambridge

*A catalogue record for this publication is available from the British Library*

*Library of Congress Cataloguing in Publication data*
Cox, Christopher.
Essentials of UMTS / Christopher Cox.
p. cm.
Includes bibliographical references and index.
ISBN 978-0-521-88931-5 (hardback)
1. Mobile communication systems. 2. Wireless communication systems.
3. Global system for mobile communications. I. Title.
TK6570.M6C69 2008
621.3845′6–dc22 2008021921

ISBN 978-0-521-88931-5 hardback

*To my Mother and Father*

# Contents

# Preface

This book is about the Universal Mobile Telecommunication System (UMTS). UMTS is the most important of the third generation (3G) mobile phone systems, which are gradually replacing the older second generation systems such as the Global System for Mobile Communications (GSM). 3G systems provide much faster communications than their predecessors, and this allows them to offer the user a wider range of services than before, such as high speed Internet access, video and interactive games.

My aim in this book has been to write a technical introduction to UMTS. As an important part of this, I have tried to give the reader a system level understanding of what all the different parts of UMTS are, and how they relate to each other. Such an understanding is hard to gain from the UMTS specifications or from the more specialised books on the subject, but is precisely what the newcomer to the system needs.

At the same time, I have kept the book short enough that it can be read cover to cover in a weekend. To do this, I have consciously left out many of the details that can be found in the specifications or in some of the other technical books on the subject. Accordingly, you won't find in this book an exhaustive description of issues such as the bit layouts in the physical channels, the contents of the system information blocks or the different types of measurement event. Rather, you will get an understanding of what those concepts are, see some examples, and gain enough knowledge to approach one of the more detailed treatments with confidence.

The book is intended for people who are new to the system, such as engineers, managers and marketing executives; it will also be valuable for those who are experienced in one part of the system but want an appreciation of what is going on elsewhere. Although it's written as a graduate level book, it assumes no previous knowledge of mobile telecommunication theory or of particular systems such as UMTS or GSM. The mathematical treatment is kept at a basic level, although an understanding of complex numbers and decibel notation will be helpful in the parts that deal with radio communications. The material goes up to the end of release 7 of the UMTS specifications, with an initial look at the issues that are being addressed in release 8.

UMTS is riddled with terminology and abbreviations, which can be a barrier to a newcomer's understanding of the subject. Although they are unavoidable, I have tried to assist the reader by putting new terms and abbreviations in italics, and by drawing attention to the terms that are particularly important for this book.

## Outline of the book

The first two chapters are introductory ones. Chapter 1 is an overview of mobile telecommunication technology, which provides the background information that will be needed by those who are new to the subject. The issues covered include radio transmission and reception, communication protocols, and the history of mobile telecommunication systems. Chapter 2 describes the system level architecture of UMTS, by looking at the hardware components that make up the system, and the software protocols that they use to communicate with each other. Its aim is to provide the reader with a framework for the later, more detailed aspects of the book.

Chapter 3 describes the techniques used for radio transmission and reception between the mobile phone and the network. The main focus is on the technology used by the air interface, which is known as wideband code division multiple access (W-CDMA). The chapter also discusses the data rates that can be reached using UMTS, and the more recent enhancements to the air interface such as high speed packet access (HSPA).

The next two chapters discuss the higher level operation of UMTS. Chapter 4 looks at the procedures that control the operation of the mobile phone, and the signalling messages that are exchanged between the mobile phone and the network. Chapter 5 then looks at the implementation of services in UMTS. It covers voice and the general packet radio service (GPRS) in some detail, and then moves on to a higher level account of other services such as the short message service (SMS).

The book concludes in Chapter 6 with a look at two technologies that are likely to be added to UMTS in the next few years: the IP multimedia system and the long term evolution of the air interface. It also describes the expected process for the introduction of fourth generation (4G) systems.

## Illustrations

Informa Telecoms & Media supplied the market research data underlying Figures 1.15 and 1.16 in Chapter 1. I am grateful to Alan Mayne and Mike Woolfrey for making the data available for use in this book.

Figures 4.5 and 4.6 have been reproduced with permission from the European Telecommunications Standards Institute (ETSI). 3GPP™ TSs and TRs are the property of ARIB, ATIS, ETSI, CCS, TTA and TTC, who jointly own the copyright in them. They are subject to further modifications and are therefore provided 'as is' for information purposes only. Further use is strictly prohibited.

# Acknowledgements

I am indebted to William Webb, joint editor of the Cambridge Wireless Essentials series, for suggesting the idea for this book and for his support and feedback while I was planning and writing it. I would also like to thank the team at Cambridge University Press, Sarah Matthews, Anna Littlewood, Eleanor Collins and Julie Lancashire, for their patience and understanding throughout the process of writing and production.

On a technical level, I am indebted to Andy Richardson for the knowledge he passed to me while delivering training courses on his behalf at Imagicom. My thanks are also due to the delegates on my training courses, for asking the questions that have stretched my understanding of the system, and for highlighting the gaps in my explanations.

Several people provided me with feedback and suggestions during the development of the book. I would particularly like to thank Stirling Essex, Julian Nolan, Mike Palmer, Rudi Tanner and William Webb, for taking time out from their Christmas holidays to review a draft of the manuscript, and for providing me with some invaluable advice on how the content and presentation could be improved. Nevertheless, the responsibility for any errors or omissions, or for any lack of clarity in the text, is entirely my own.

# 1 Introduction to mobile telecommunications

Mobile phones were first introduced in the early 1980s. In the succeeding years, the underlying technology has gone through three phases, known as generations. The first generation (1G) phones used analogue communication techniques: they were bulky and expensive, and were regarded as luxury items. Mobile phones only became widely used from the mid 1990s, with the introduction of second generation (2G) technologies such as the *Global System for Mobile Communications* (GSM). These use more powerful digital communication techniques, which have allowed their cost to plummet, and have also allowed them to provide a wider range of services than before. Examples include text messaging, email and basic access to the Internet.

Third generation (3G) phones still use digital communications, but they send and receive their signals in a very different way from their predecessors. This allows them to support much higher data rates than before, and hence to provide more demanding services such as video calls and high speed Internet access. This book is about the most popular third generation technology, the *Universal Mobile Telecommunication System* (UMTS).

The first chapter lays the foundations for the subjects covered later in the book. It begins by briefly describing the architecture of a mobile telecommunication system, and continues with a more detailed look at two important aspects of its operation: the communication protocols that manage the delivery of information to and from a mobile phone, and the special techniques that are used for radio transmission and reception. It concludes with the history of mobile telecommunications, and a look at the current state of the market.

## 1.1 Architecture of a mobile telecommunication system

Figure 1.1 shows the architecture of a mobile telecommunication system. The system is controlled by a particular network operator such

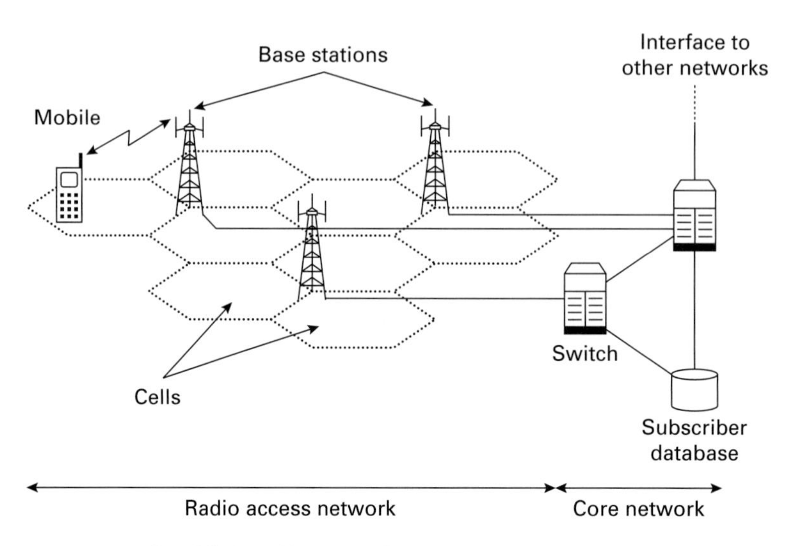


**Figure 1.1** Simplified architecture of a mobile telecommunication system.

as Vodafone or O2, and is often known as a *public land mobile network* (PLMN). It has three main components: the core network, the radio access network and the mobile phone.

The *core network* has a similar role to a traditional fixed line telephone network. It sends information like voice calls or text messages from one phone to another using components that are known as *switches*. It also maintains a database containing information about the network operator's subscribers, and uses the database for tasks like preparing and distributing bills. Finally, it has a number of functions that are not required in a fixed line network; for example, it monitors the locations of the mobile phones, so that it can continue sending information to them as they move around.

The *radio access network* handles the radio communications between the core network and the mobile phone. It contains a large number of *base stations*, each of which transmits and receives radio signals to and from the mobile phones in the surrounding area. The area around a base station is often divided into multiple *sectors* by equipping the base station with multiple directional antennas, each of which communicates with the mobile phones in the corresponding sector.

If this is done, then the number of sectors per base station is usually either three (as shown in the figure) or two.

The most common word for a part of the radio access network is *cell*. Unfortunately this word is ambiguous: it can refer either to a single sector, or to the group of sectors that a base station controls. We will use the first definition throughout this book, so that the words ‘cell’ and ‘sector’ will mean exactly the same thing.

Each cell has a maximum size, which is determined by the greatest distance at which the radio signals can be successfully received. It also has a maximum capacity, which limits the number of mobile phones that can make calls within the cell at the same time. In rural areas, the population densities are low, so capacity is not a problem. The cells are therefore large, typically a few kilometres across, and are known as *macrocells*. In urban areas, the population densities are much greater. To handle this, we can introduce an extra set of *microcells*: these are only a few hundred metres across, so they greatly increase the total capacity of the network. The original macrocells are usually retained, as they are useful for fast-moving users who move quickly from one cell to another. We can also use a third set of *picocells*: these are a few tens of metres across, and provide small-scale coverage in offices, shopping centres or the home.

The use of cells is a crucial part of the system: it allows the same radio frequencies to be used in different locations with little interference, which greatly increases the number of mobile phones that can be supported. For this reason, the system is often known as a *mobile cellular network*.

The user’s device was traditionally known as a mobile phone but, with the increased use of data communications like text messaging and email, this terminology has become rather restrictive. In UMTS, the device is officially known as the *user equipment* (UE); in this book, we normally use the simple term *mobile*. The interface between the radio access network and the mobile is known as the *air interface* or the *radio interface*. On this interface, the path from the network to the mobile is known as the *downlink* (DL) or *forward link*, and the path from mobile to network is the *uplink* (UL) or *reverse link*.

When a mobile moves from one cell to another, it has to stop communicating with its current cell and start communicating with the next cell along. This process is known as a *handover*, and is controlled by signalling messages between the mobile and the network. A mobile can also move outside the region covered by its own network operator, for example when travelling to another country. The mobile can still make calls by using resources in two networks: the base stations in the *visited network*, the user database in the *home network*, and switches in both. This situation is known as *roaming*.

There are several books with more information about mobile telecommunication systems: reference [1] is an excellent example. In the next couple of sections, we will examine two aspects of the system in more detail: the communication software that transfers data and signalling messages across the network, and the techniques that are used for reliable transmission and reception over the air interface.

## 1.2 Communication networks

The role of a communication network is to connect devices like computers and phones to each other so that they can exchange data and signalling messages. The network has to accept information from a transmitting device, identify a route to the receiver, and send the information there without introducing any significant errors. Some example communication networks include fixed line telephone systems, the Internet, and the mobile communication network that we introduced in Figure 1.1.

This section is an introduction to the tasks that communication networks carry out, with some examples of the communication software that they use. There are many books that give detailed accounts, such as references [2], [3] and [4].

### 1.2.1 Circuit switching and packet switching

Communication networks can transport information using two very different techniques, which will both be important for this book: circuit switching and packet switching. The two techniques are illustrated in Figure 1.2. Circuit switching is generally used for voice calls, and packet switching for data.

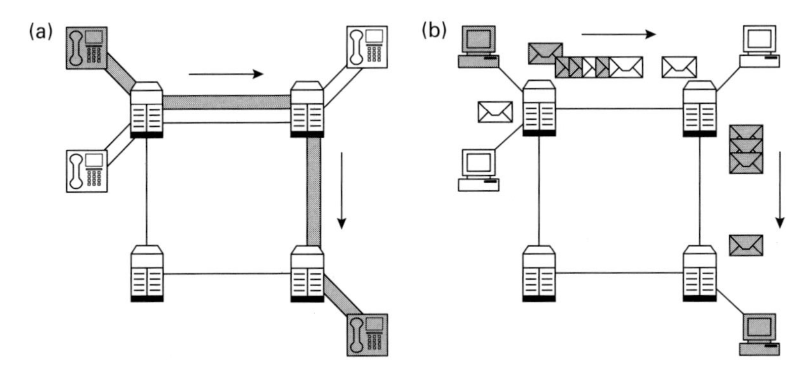


**Figure 1.2** Illustration of the two main transport mechanisms in a communication network. (a) Circuit switching. (b) Packet switching.

*Circuit switched* (CS) networks (Figure 1.2a) use the same techniques as a traditional fixed line telephone system. At the start of a call, the network identifies a route through the switches that connect the two phones, and reserves enough resources on that route to handle the call. For example, a voice call typically requires a constant data rate of 64 000 bits per second (64 kbps). By reserving enough resources in the switches and the intervening links for transmission at 64 kbps, we can ensure that the information travels from end to end with a very low delay and with no obstruction from other calls.

Circuit switching has a big disadvantage, however: it is rather inefficient. In a phone call, each user is only speaking for half the time on average, so we have already set aside twice the resource that we actually need. The situation is worse when doing data transfers such as web browsing because these are typically very bursty, comprising short periods of activity separated by long periods when nothing is happening.

To deal with this problem, *packet switched* (PS) networks like the Internet use a different technique (Figure 1.2b). In this technique, the transmitter divides the data stream into blocks that are known as packets. It adds some extra information, known as a header, to each packet, which tells the network how the packet should be routed. It then sends each successive packet to the first switch in the network. When a packet reaches a switch, the switch looks up the packet's routing information in a routing table, reads the identity of the next switch in the route, and forwards it there. This process is repeated switch by switch, so that eventually the packet reaches its destination.

The packets can be routed in two different ways. In the *virtual circuit* approach, each packet is labelled with a virtual circuit number that identifies the data stream. The routing table lists the virtual circuits that a particular switch is using, and identifies the next switch in the route for each one. *Permanent virtual circuits* last indefinitely, while *temporary virtual circuits* are set up for the duration of a single data stream. Temporary virtual circuits are more flexible, but they require some initial signalling at the time the data stream is set up, to identify a route through the network and set up the routing tables. The *datagram* approach is rather different. In this approach, each packet is labelled with the address of the destination device. The routing table lists all the destination addresses that the switch might have to use, and maps each destination address onto the identity of the next switch in the route. The datagram approach is the more popular of the two (it is used for routing in the Internet, for example), but we will see both approaches in this book.

Packet switching is more efficient than circuit switching, because if a source stops transmitting, then its resources are immediately available for other data streams. It has a disadvantage, however: if several sources decide to transmit at once, then their total data rate can exceed the capacity of the intervening links, and the network can become congested. If this happens, then the packets are held in queues in the network's switches, which leads to delays.

Because packet switched networks are more efficient than circuit switched ones, there is currently a trend among telecommunication operators to work round the problems noted above, and to introduce packet switched transport for all services, voice as well as data. (The use of packet switched networks for voice calls is often known as *voice over internet protocol* (VoIP).) We will see this trend reflected at various points in the book.

### 1.2.2 Communication protocols

Routing is just one of the functions of a communication network. Other functions include controlling the electrical signals on each interface, encrypting the information if it has to be transmitted securely, and possibly retransmitting the information if an error occurs. To keep these functions separate, each of them is handled by a software component known as a *protocol*, and the individual protocols are arranged into a *stack* that has several different *layers*. In the transmitter, the information is processed first by the higher layer protocols and then by the lower layer ones, before sending it into the communication network. The process is reversed in the receiver, to recover the original information.

There are different ways to arrange the layers in a protocol stack, but the most common is the seven-layer OSI (*open systems interconnection*) model shown in Figure 1.3. The figure just shows the processes in the transmitter and the receiver: we will cover what happens inside the network in a few moments. The stack will be described by reference to a packet switched network, although many of the issues apply to a circuit switched network as well.

Above the protocol stack, the application software is something like a web browser or an email client. The *application layer* (layer 7) acts as an

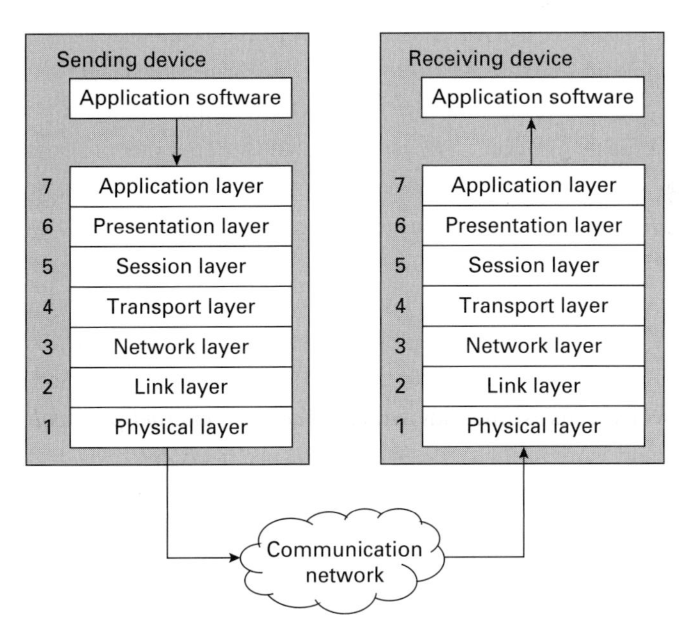


**Figure 1.3** Organisation of communication protocols in the OSI protocol stack.

interface between the application and the lower layer protocols, by providing software functions for tasks such as setting up a data stream and sending a data packet. Some well-known application layer protocols are the *hypertext transfer protocol* (HTTP) and the *simple mail transfer protocol* (SMTP), which handle web pages and emails respectively. The *presentation layer* (6) represents the information being exchanged between the two end devices using a common syntax that both can understand, while the *session layer* (5) sets up, manages and tears down the communication path between them. Layers 5 and 6 are less important than the others, and we will not consider them further.

The *transport layer* (4) manages end-to-end transfers from the transmitter to the receiver, without worrying about the intervening route. There are two main types of transport layer protocol. Connection-oriented protocols, like the *transmission control protocol* (TCP), use signalling communications between the transmitter and receiver as well as the actual data transfer. This brings a number of benefits, for example it allows the receiver to request retransmissions of data that have arrived incorrectly. However, it also slows the data transfer down. Connectionless protocols, like the *user datagram protocol* (UDP), just send data to the receiver without any extra signalling. They are suitable for information like streaming video, for which timely arrival is more important than perfect accuracy.

The *network layer* (3) ensures that data are sent on the correct route from transmitter to receiver. The network layer protocol used on the Internet is the *Internet protocol* (IP), which uses the datagram approach and carries out routing using the IP address of the destination device. The *link layer* (2) sends data on a single link from one switch to another. Like the transport layer, the link layer can be connection-oriented or connectionless: the difference is that any layer 2 retransmissions are on a link-by-link basis, while layer 4 retransmissions are made end-to-end. Two common link layer protocols are *Ethernet* and the *point-to-point protocol* (PPP). The link layer also manages the underlying *physical layer* (1): this transmits and receives the actual signals, using a transmission medium such as copper wire, optical fibre or radio.

We can think of the interactions between different layers in two ways. These are shown in Figure 1.4, using the link layer as an example. The

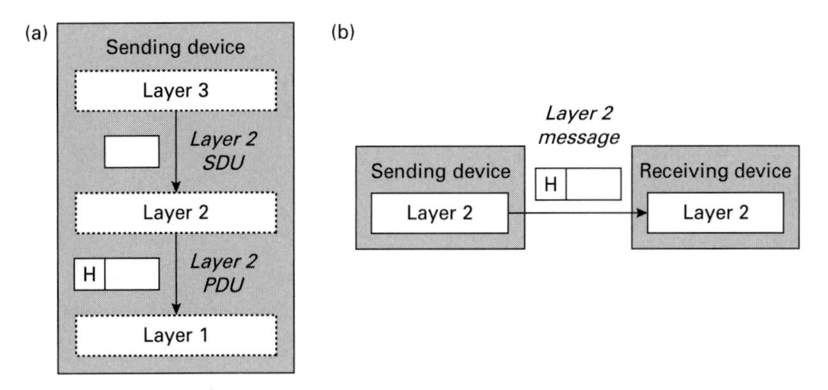


**Figure 1.4** Illustration of how data packets are transferred between communication protocols. (a) Transfer between the protocol layers in a single device. (b) Transfer between two different devices.

first way is to think of the vertical interactions inside a single device (Figure 1.4a). In the transmitter, layer 3 sends a packet to layer 2, which is known as a layer 2 *service data unit* (SDU). Layer 2 processes the packet and adds a header, denoted H in the figure. (The processing might include link-by-link encryption, for example, in which case the header might indicate that encryption has been used.) It then sends the new packet down to layer 1 for further processing. The new packet is known as a layer 2 *protocol data unit* (PDU), although it immediately becomes a layer 1 SDU. In the receiver, layer 2 receives a PDU from layer 1. It detaches and inspects the header, undoes the effect of the transmit processing (here by decryption), and passes the resultant SDU up to layer 3.

The second way (Figure 1.4b) is to think of the horizontal interactions between devices. From this point of view, the transmitter's link layer sends a message to the receiver's link layer, which contains the header and the processed data. It uses the layer 1 protocol to do this, but the details of that protocol are hidden from the link layer and can be thought of as a black box. The effect is that the details of each layer can be isolated from the other layers in the protocol stack.

What happens inside an individual switch? A typical answer is shown in Figure 1.5. In this figure, the switch is receiving packets on an Ethernet link from the source device, and has to send them to the

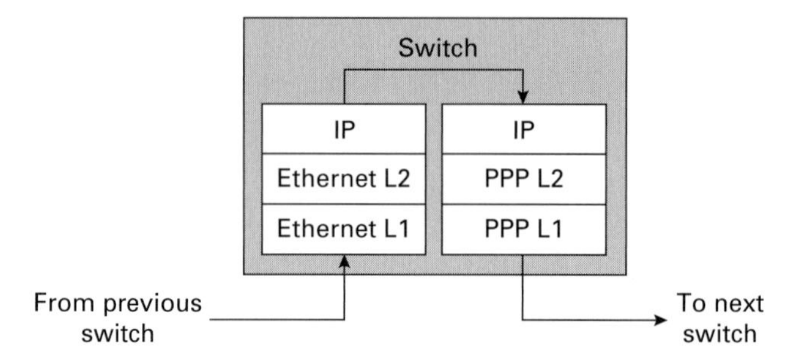


**Figure 1.5** Example operation of the layer 1, 2 and 3 protocols inside a switch.

destination device using PPP. To do this, it unwraps the received packets as far as the layer 3 PDU, reads the routing information there, and makes any changes that are needed to the layer 3 header. It then wraps the packets up again using PPP, and sends them on the correct route to the destination device. In a packet switched network, a layer 3 switch like the one in the figure is often known as a *router*.

In a complex system like UMTS, the individual layers are often subdivided into more than one protocol, each of which handles one aspect of that layer's functions. In addition, UMTS needs a lot of extra signalling to carry out tasks such as roaming and handover. These signalling functions are often handled by a separate protocol stack, which we can think of as an extra version of Figure 1.3. If this is done, then the signalling functions are collectively known as the *control plane*, while the data transfer part is known as the *user plane*.

### 1.2.3 Example communication protocols

To illustrate the points discussed above, Figure 1.6 shows three protocol stacks that we will use later on in the book: the Internet protocol stack, ATM and SS7.

The Internet protocol stack uses several protocols that we have already introduced, such as TCP, UDP and IP. It carries out routing by datagram-based packet switching, while layers 1 and 2 can use any mechanism at all, such as Ethernet.

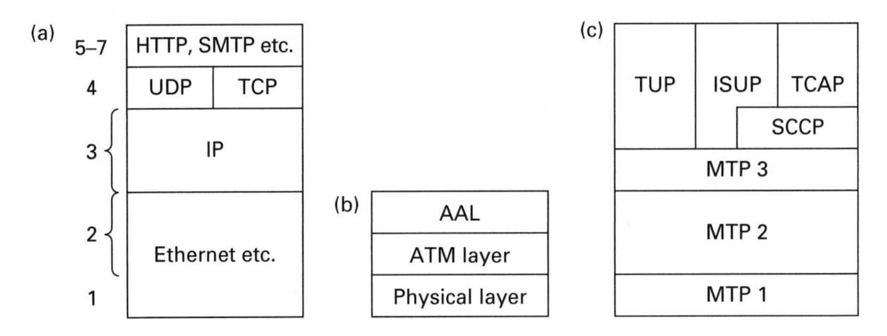


**Figure 1.6** Example protocol stacks. (a) Internet, (b) ATM, (c) SS7.

*Asynchronous transfer mode* (ATM) is another packet switched protocol stack, but one that uses the virtual circuit approach. It was designed for the high speed, end-to-end transfer of both real time and non-real time information streams, but its main use nowadays is in high speed network backbones like the core and radio access networks of UMTS. When implemented in this way, it has three layers that occupy layers 1 and 2 of the OSI stack: the physical layer, the ATM layer and the ATM adaptation layer (AAL).

*Signalling system 7* (SS7) is a control plane protocol stack that handles signalling messages in fixed line telephone networks all over the world. The main application layer protocol is known as the *ISDN user part* (ISUP): this contains all the signalling messages that are required by a digital telephone network, such as messages for setting up a call, modifying it and tearing it down. Some older networks still use analogue transport, and they require an older application layer protocol known as the *telephone user part* (TUP). A third protocol known as the *transaction capabilities application part* (TCAP) acts as an interface to other application layers, so that devices can exchange messages that are not defined by ISUP or TUP. The *message transfer part* (MTP) handles transport, while the *signalling connection control part* (SCCP) improves the routing capabilities of MTP.

These protocol stacks can be combined. For example, ATM can be used for layer 2 transport in an IP network, and SS7 messages can also be transported using IP, ATM, or IP over ATM. Approaches like these are used in several parts of UMTS, but they cause complications like the need for extra protocols to act as software interfaces, so we will not describe the details.

## 1.3 Digital wireless communications

We will now look at a particularly important issue in a digital communication system. How do we send information from the transmitter to the receiver as quickly as possible, without introducing any errors into the information stream?

Figure 1.7 shows the processes that are involved. In mobile telecommunications, it turns out that these processes are especially important on the air interface, because the capacity of the air interface is often the factor that limits the total capacity of the system. The processes will therefore be described by reference to the air interface, even though most of them can be applied to other transmission media as well, such as copper wire or optical fibre. There are several books to consult for more information, notably [5].

### 1.3.1 Modulation

We begin with the last component in the transmitter, the *modulator*. The input to the modulator is a stream of binary 1s and 0s, which are the bits

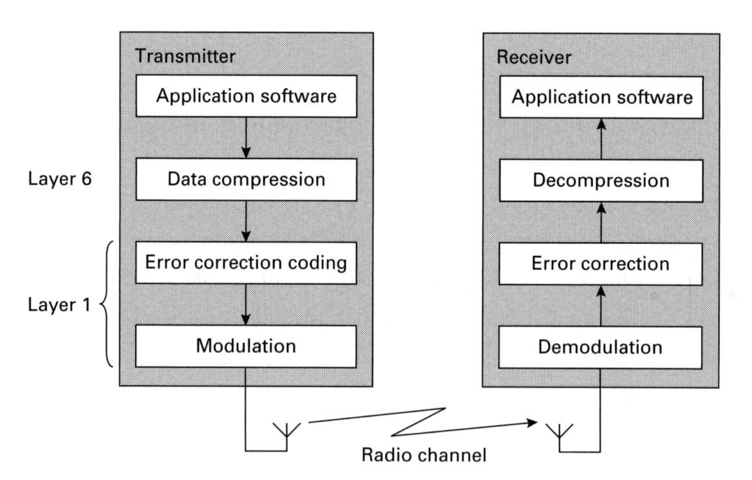


**Figure 1.7** Illustration of the most important processes in a digital wireless communication system.

that we want to send. The modulator first converts these bits into *symbols*, each of which is a signal that represents either one bit or several bits in parallel. It uses the symbols to modify a sinusoidal waveform known as a *carrier*, and it then sends the modified waveform to the receiver.

Figure 1.8 shows two simple modulation schemes, which are both used in UMTS: *binary phase shift keying* (BPSK) and *quadrature phase shift keying* (QPSK). In BPSK, the modulator maps bits of 0 and 1 onto symbols of $+1$ and $-1$ respectively. Figure 1.8a shows these symbols as points on a number line, in a diagram known as a constellation. The modulator then multiplies the symbols by a sine wave carrier, to produce the waveform shown in Figure 1.8b. Inspection of the waveform shows that we can actually think of BPSK in two different ways: as either an amplitude modulation using symbol amplitudes of $+1$ and $-1$, or a phase modulation using symbol phases of 0° and 180°.

QPSK (Figures 1.8c and 1.8d) works in a similar way, but it takes bits two at a time and maps them onto four different symbols, with phases of 45°, 135°, 225° and 315°. We can think of QPSK as an amplitude modulation too, but the symbols are now complex numbers with values of $\pm 1 \pm i$, where i is the square root of $-1$. For example, a bit sequence of 01 is represented using a symbol of $1 - i$, in which the real part is $+1$ and the imaginary part is $-1$. The real and imaginary parts are often known as the *in-phase* (I) and *quadrature* (Q) components of the signal.

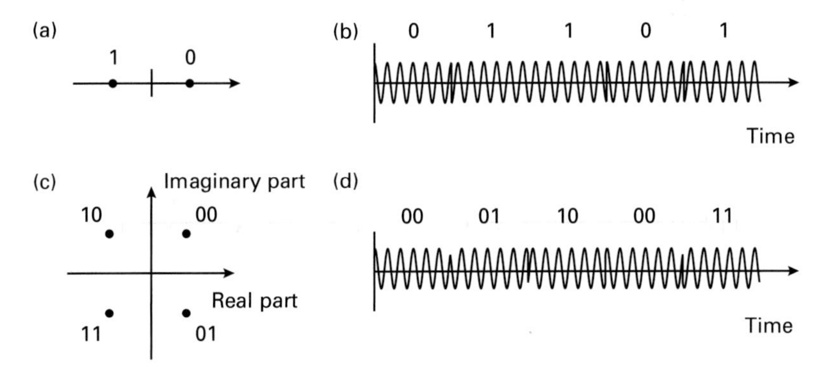


**Figure 1.8** Constellation diagrams and signal waveforms for two common modulation schemes. (a) Constellation diagram for BPSK. (b) Example BPSK waveform. (c) Constellation diagram for QPSK. (d) Example QPSK waveform.

### 1.3.2 Multiple access

The modulation schemes described above are fine for one-to-one communications, but in a mobile cellular network each base station has to transmit data to several mobiles at the same time, and receive data back from them. We need a way to distinguish those data streams so that (on the downlink, for example) each mobile can process the data that have been sent to it, and ignore the data intended for the other mobiles in the cell.

The problem is solved using multiple access, for which there are a number of different techniques (Figure 1.9). In *frequency division multiple access* (FDMA), the base station transmits to each mobile on a different carrier frequency, and the mobiles select their own signals using analogue filters. In *time division multiple access* (TDMA), the base station sends short bursts of data to each mobile in succession, and the mobiles select their own bursts by receiving signals only at the required times.

UMTS uses a different technique, known as *code division multiple access* (CDMA). In CDMA, the base station transmits to all its mobiles at the same time and on the same carrier frequency. However, it uses a modified modulation scheme, in which each data stream is labelled with a code that is unique to the destination mobile. By processing the received signal using its own code, a mobile can extract the data that are intended for it, and discard the data intended for all the other mobiles in the cell.

CDMA is more complex than FDMA or TDMA, but it allows each cell to support a larger number of mobiles than before. We will discuss the details of CDMA in Chapter 3, together with the advantages and difficulties that it brings.

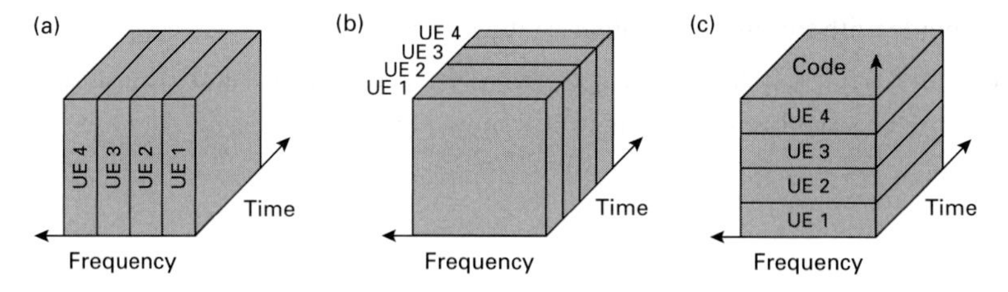


**Figure 1.9** Illustration of three common multiple access techniques. (a) FDMA, (b) TDMA, (c) CDMA.

### 1.3.3 Radio channel

We now move to the bottom of Figure 1.7, and look at the radio channel. This is the transmission medium between the transmitter and the receiver. It modifies and distorts the transmitted signal due to propagation loss, noise and fading.

*Propagation loss* is the reduction in signal power as the transmitted waveform spreads out in space. If there were no obstacles between transmitter and receiver, then the received signal power would follow an inverse square law and be proportional to $1/r^2$, where $r$ is the distance between the two. The situation in a mobile cellular network is more complicated, due to issues like reflection of rays off the ground and diffraction around buildings. In macrocells, we find that the average received power is roughly proportional to $1/r^m$, where $m$ lies between about 3.5 and 4. Radio propagation in micro- and picocells is sensitive to the exact sizes and locations of the individual obstructions, and usually has to be calculated on a ray-by-ray basis for each individual cell.

The received signal is also degraded by noise, which has two main causes: thermal noise in the receiver and interference from other transmitters. Noise makes the received signal fluctuate about its average value, which can lead to bit errors in the receiver if (for example) a transmitted symbol of $+1$ is misinterpreted as a $-1$.

The third problem is *fading*. The transmitted signal usually travels to the receiver over several different ray paths, in a phenomenon known as *multipath* (Figure 1.10a). At the receive antenna, the rays add together and interfere. This interference can be constructive if the peaks of the received waveforms coincide, or it can be destructive if the peaks on one ray coincide with troughs on another. If the mobile or the reflectors move, then the interference pattern changes between constructive and destructive, and the received power fluctuates about its average value (Figure 1.10b), occasionally reaching very low values that are known as *fades*.

The exact fading pattern depends on the ray geometry, but a common model is Rayleigh fading, in which we assume that the rays reach the receiver in the horizontal plane and equally from all angles. In Rayleigh fading, the distance between fades is roughly half a wavelength of the

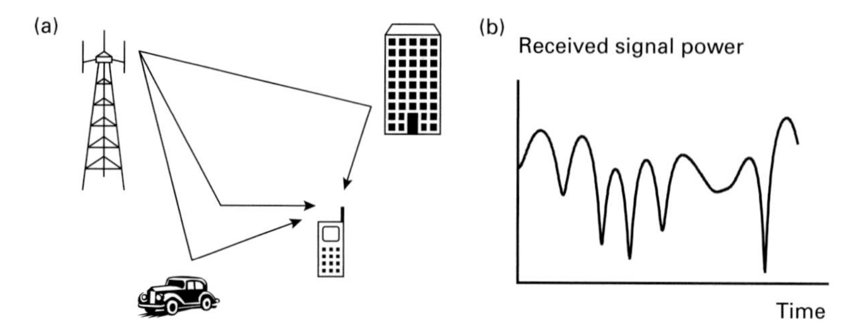


**Figure 1.10** Illustration of multipath and fading. (a) Multipath propagation from a base station transmitter to a mobile receiver. (b) Example fading pattern, showing how the received signal power varies as the mobile or reflectors move.

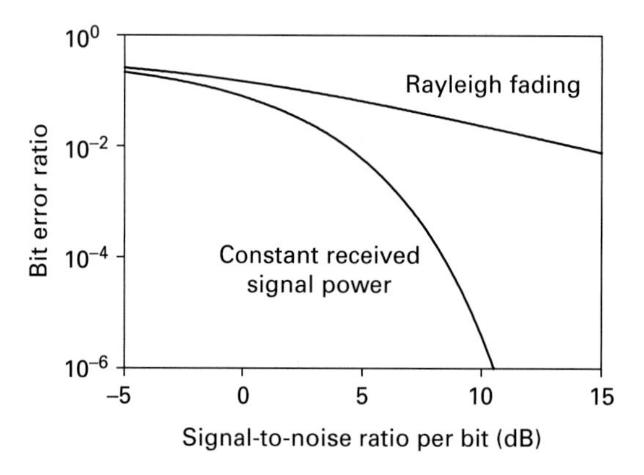


**Figure 1.11** Bit error ratios for BPSK in the absence of error correction, for the cases of a constant received signal power and Rayleigh fading.

carrier signal. In UMTS, the carrier frequency is usually about 2 GHz, so the fades are about 50 to 100 mm apart.

In some situations, we can predict the number of bit errors in the received signal. Figure 1.11 shows the probability of a bit error as a function of a quantity denoted $E_b/N_0$, which expresses the signal-to-noise ratio per bit. (The $x$-axis is plotted in decibels, so the quantity shown is actually $10 \log_{10} (E_b/N_0)$.) The figure shows two situations: a constant signal power and Rayleigh fading. It assumes that the modulation scheme is BPSK, and that no further steps have been taken to correct the errors. As the signal-to-noise ratio increases, so the probability of a bit error falls, but the error rate is much greater in the case of a Rayleigh fading signal. This illustrates an important fact: fading signals are much harder to process than signals with constant amplitude, even if the average power received is the same. Mobile telecommunication systems use several techniques to reduce the amount of fading, and we will now examine one of these.

### 1.3.4 Diversity processing

In *diversity processing*, the receiver picks up multiple copies of the same signal, and combines them to produce a signal with more power and less fading. An example (Figure 1.12) is receive antenna diversity, in which a base station uses two receive antennas that are typically spaced a metre or two apart.

Receive antenna diversity brings two benefits. The first and more obvious one is that it increases the received signal power, but that isn't all. If the two antennas are far enough apart, then the ray paths from the transmitter to the two antennas will be very different, and the two received signals will undergo fades at different times. By processing the two signals independently to ensure that they don't interfere, we can add the signal powers received on the two rays, and reduce the total amount of fading. Both the increase in power and the reduction in fading serve to reduce the error rate.

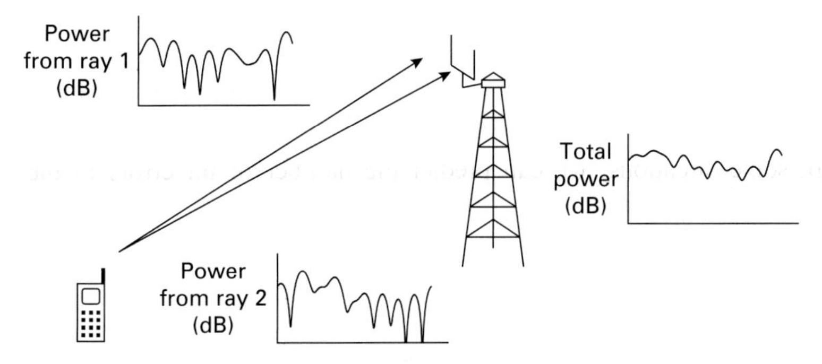


**Figure 1.12** Reduction in fading by the use of receive antenna diversity processing.

Receive antenna diversity is just one example of diversity processing. Some other common examples are polarisation diversity (the reception of two different polarisations), and transmit antenna diversity (the transmission of a signal from multiple antennas). UMTS base stations normally use two-antenna receive diversity on the uplink: they could use more, but the extra benefits would be small, and the masts would be even larger and uglier than they already are. Optionally, they can also use two-antenna transmit diversity on the downlink, but this is less common as the technique is more complex and requires more transmit power. We will see two other examples of diversity processing in UMTS later on: soft handover (Section 2.2.3) and the rake receiver (Section 3.1.5).

### 1.3.5 Error correction

A simple reading of Figure 1.11 suggests that we need very high signal-to-noise ratios for reliable digital communication, particularly in the presence of fading. Fortunately, the situation is not as bad as it first appears.

The solution lies in the *error correction* coder that we introduced in Figure 1.7. This represents the information we want to transmit using codewords that typically have two or three times as many bits. For example, it might represent the three information bits 101 using the six bit codeword 111010. If an error happens, then the receiver picks up a different codeword such as 011010. If the number of errors is small enough and the coder is well designed, then the received codeword will not correspond to any possible sequence of information bits. Furthermore, the receiver will be able to conclude that the closest set of information bits was 101, and that an error happened in the first bit of the codeword. The process is similar to a written language like English, in which the individual letters supply extra data that allow the reader to understand the underlying meaning, even if sdme of jhe wodds are mis-spult.

Error correction is even more powerful than it first appears. In 1948, Claude Shannon [6] proved that it was theoretically possible to send digital signals over a communication link without any errors at all, provided that the information rate lay below a maximum value, known as the channel capacity. In the absence of fading, we can calculate the

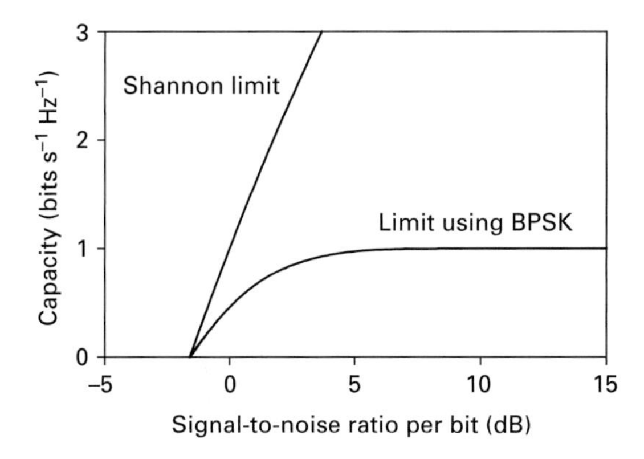


**Figure 1.13** Channel capacity in the absence of fading, for the cases of BPSK and an arbitrary modulation scheme.

channel capacity as follows:

$$\frac{C}{W} = \log_2\left(1 + \frac{C}{W}\frac{E_b}{N_0}\right) \tag{1.1}$$

where $C$ is the channel capacity, $W$ is the signal bandwidth, and $E_b/N_0$ is the signal-to-noise ratio per bit. Figure 1.13 illustrates this, by showing the theoretical channel capacity as a function of $E_b/N_0$. It also shows the channel capacity if the modulation scheme is restricted to BPSK: this is less than the maximum value, but it can usefully be contrasted with the error rate without fading that we showed in Figure 1.11.

The word 'theoretically' hides many practical problems: in particular, it requires the use of a perfect error correction algorithm, which almost by definition does not exist. However, many of the recent advances in communication technology can be seen as attempts to get ever closer to the limits implied by Shannon's equation. Two examples of these are the introduction of turbo coders in UMTS (Section 3.1.6) and high speed packet access (HSPA) (Section 3.2).

### 1.3.6 Data compression

Before the transmitted data stream reaches the physical layer, it is usually compressed to reduce its data rate while preserving all the useful information, or as much of it as we can. This process increases the number of information streams that can be successfully carried within the channel capacity *C*. Data compression is particularly important on the air interface, which is normally the main bottleneck in a mobile communication system. For example, speech is usually sampled at a rate of 64 kbps, but in UMTS it is compressed to a rate between 4.75 and 12.2 kbps before transmission.

It is useful to close this section with a brief comparison of analogue and digital communication technologies. Digital communication has long superseded analogue in the field of mobile phones, and is well on the way to doing so in radio and television. There is nothing magical about it, however: it simply lets us manipulate the transmitted and received signals much more easily than before. In particular, data compression ensures that we only need to transmit the information that is actually important, while error correction ensures that the data reach their destination correctly. Together, they greatly increase the amount of information that we can transmit in a given frequency band.

## 1.4 History of mobile telecommunication systems

We close this chapter by describing how mobile telecommunication systems have evolved in the years since they were first introduced. This section is less important to us than the earlier ones, but it will provide some useful background information for those who are new to the subject.

The history of mobile telecommunication systems is depicted in Figure 1.14. Each system is shown from the year in which it was first deployed: in the case of older systems, the dashed lines on the right denote the period between the shutdown of the first network and the final one. Branching lines connect older systems that have evolved into younger ones, such as GSM and UMTS. The systems are often grouped into three generations, which we will now describe.

### 1.4.1 First generation

*First generation* (1G) mobile cellular networks were introduced in the early 1980s. They were all based on frequency division multiple access

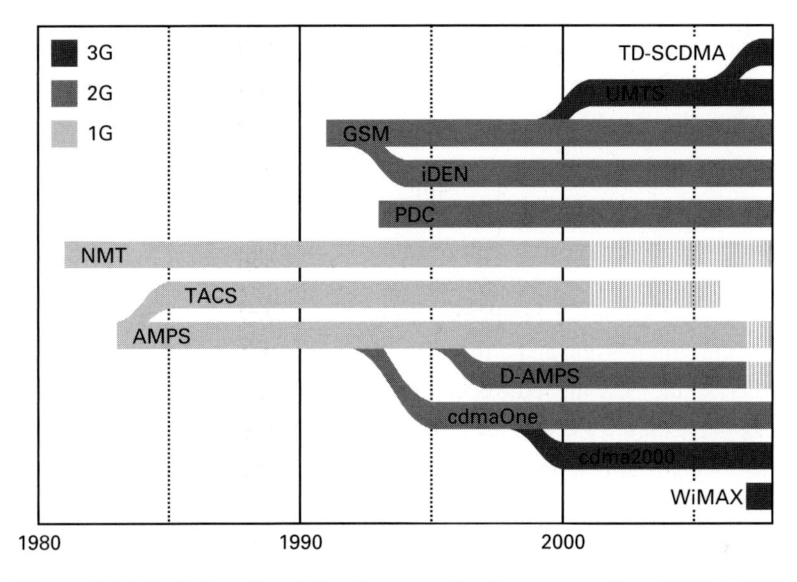


**Figure 1.14** Evolution of mobile telecommunication systems from 1980 to 2008.

and analogue signal processing, but had most of the important attributes of today's systems, notably handover and in some cases international roaming.

1G systems were typically designed for use in individual countries, but three of them were especially successful. *Nordic Mobile Telephone* (NMT) was developed in Scandinavia but was adopted by several other European countries, while the *Advanced Mobile Phone System* (AMPS) was developed in the USA and exported mainly to the Far East. The *Total Access Communication System* (TACS) was a UK derivative of AMPS, and was used in geographical regions that overlapped those of AMPS and NMT.

1G systems had a couple of disadvantages. They used large cells (several kilometres across), so they had much lower capacities than today's networks, and this inevitably kept costs to the subscriber high. They also had little security protection, and this led to several well-publicised problems: with the right equipment, intruders could eavesdrop on phone conversations, and could clone handsets using the information gained from signalling messages. First generation networks have been in decline for many years as users moved over to more up-to-date systems. Several networks have already shut down, and some of the frequency spectrum has been re-allocated.

### 1.4.2 Second generation

First generation systems used analogue communication technology, but *second generation* (2G) systems are digital. This allows them to implement the compression and error correction technologies that we described earlier, so they use the radio spectrum much more efficiently than before. Together with the use of smaller cells, this means that second generation networks have much higher capacities than first generation ones.

The *Global System for Mobile Communications* (GSM) is the most important. It was conceived in the 1980s as a system that could be used throughout Europe, and was first deployed in 1991. However, it has been so successful that it is now the dominant system almost everywhere, and has around 80 per cent of the world's subscribers. GSM uses a mix of FDMA and TDMA technologies: each cell has a number of different carrier frequencies, each of which is divided into eight timeslots. It has more powerful security features than 1G systems, including (for example) the encryption of data and signalling messages on the path between the mobile phone and the base station. Another technology first introduced in GSM is the *subscriber identity module* (SIM), a smart card which contains the subscriber's personal details and which can be moved from one handset to another.

GSM was originally designed just for voice calls, but has had several enhancements. A notable example is the *short message service* (SMS), which was introduced in the mid 1990s, and whose success took network operators by surprise. In 2006, data accounted for 13.5 per cent of worldwide revenues [7], but the vast majority of this was from SMS.

Later enhancements to GSM are often known as 2.5G systems, and have two main themes: higher bit rates, and the delivery of data as well as voice. The *general packet radio service* (GPRS) was designed for the efficient delivery of data to mobile devices. It required the construction of a packet switched core network, to run alongside the circuit switched network that was originally built for GSM. *Enhanced Data Rates for Global Evolution* (EDGE) modifies the modulation scheme used by GSM, so as to increase the bit rate per user and the capacity of the system. EDGE is popular in North America, where the allocation of carrier frequencies has made it hard for GSM operators to upgrade to UMTS.

The other main 2G system is *cdmaOne*, also known as *IS-95* from its interim standard number. It was designed by Qualcomm in the early 1990s and deployed from 1995; it is now the most popular system in the USA and Canada, and has also been widely adopted in the Far East. As its name implies, cdmaOne was the first mobile cellular system to use CDMA technology. Each cell uses a carrier with a bandwidth of 1.25 MHz, which is divided into 64 data and signalling channels by the use of codes. In the core network it uses a specification known as IS-41, which was inherited from AMPS. As with GSM, the IS-95 specifications have had a number of enhancements. The most commonly used is IS-95B, which supports packet data at rates up to 64 kbps, and which is usually classified as a 2.5G system in a similar way to GPRS.

It is also worth noting three other 2G systems. *Digital AMPS* (D-AMPS, otherwise known as US-TDMA and by the standards IS-54 and IS-136) is an enhancement to AMPS, in which the network's capacity was increased by the use of digital technology and by dividing each carrier into three timeslots. The *Integrated Digital Enhanced Network* (iDEN) is a proprietary system designed by Motorola, which is based on GSM but has been modified to support other services such as two-way radio. Finally, *Personal Digital Cellular* (PDC) is another TDMA-based system, which was developed and widely used in Japan, but is now being phased out in favour of third generation systems.

### 1.4.3 Third generation

*Third generation* (3G) systems are characterised by changes to the air interface that support higher bit rates. 3G systems were first introduced from the late 1990s in a process led by the *International Telecommunication Union* (ITU), the United Nations agency for communications and information technology. In 1997, the ITU published a set of requirements for 3G communication technologies under the name of IMT-2000 (*International Mobile Telecommunications 2000*). Several proposals were submitted and, in 1999, the ITU selected five systems as being compatible with IMT-2000. After some years of evolution, four main systems remain.

The subject of this book is the *Universal Mobile Telecommunication System* (UMTS). This system was developed from GSM, by keeping the core network more-or-less intact but changing the air interface to use CDMA. There is some compatibility between the two systems: most UMTS mobiles also implement GSM, and the network can hand them over from a UMTS base station to a GSM one if they reach the edge of the UMTS coverage area. However, network operators cannot implement the two systems in the same frequency band, so they are not fully compatible with each other.

The air interface of UMTS is based on the same principles as that of cdmaOne, but the details are very different. In particular, it uses a higher signal bandwidth of 5 MHz, so is often known as *wideband CDMA* (W-CDMA). There are two modes of operation, known as *frequency division duplex* (FDD) and *time division duplex* (TDD). Of these, FDD is currently much more popular, to the extent that we will hardly discuss TDD at all.

Since it was originally introduced, UMTS has been upgraded in a similar way to GSM. The best known of the new technologies is *high speed packet access* (HSPA), which has been designed to increase the bit rates on the air interface in the case of non-real time packet data. By analogy with GPRS and EDGE, HSPA is generally described as a 3.5G technology.

*cdma2000* is a direct upgrade of cdmaOne: the two systems can co-exist on the same carrier frequency, with cdmaOne mobiles communicating successfully with cdma2000 base stations and vice-versa. This is a smoother upgrade path than the one from GSM to UMTS, and has aided the adoption of cdma2000 in North America. There are two main variants. cdma2000 1xRTT (*1x Radio Transmission Technology*) has twice the capacity of cdmaOne, which is achieved by introducing 64 more codes to run alongside the original set. In cdma2000 1xEV-DO (*1x Evolution Data Optimised*), separate 1.25 MHz carriers are used for voice and data. The latter are then optimised for the delivery of data, with techniques similar to those used by HSPA.

The name of the third 3G system is rather a mouthful: *time division synchronous code division multiple access* (TD-SCDMA). This is actually a variant of UMTS, also known as the *TDD low chip rate option*. TD-SCDMA is being developed in China, with a view to promoting Chinese technology and minimising the royalties paid to Europe and the USA. At the time of writing, TD-SCDMA trials were under way, with the first services expected in time for the 2008 Olympics. However, licences are unlikely to be issued until later, perhaps in 2009.

In 2007, the ITU approved another system as compatible with IMT-2000: *Worldwide Interoperability for Microwave Access* (WiMAX). This system uses a different technology from earlier 3G systems, known as *orthogonal frequency division multiple access* (OFDMA), and has evolved in a very different way. The original specification, IEEE 802.16-2001, was nothing like a cellular technology: instead, it was designed for line-of-sight communications between stationary devices. IEEE 802.16-2004 (802.16d) added support for non-line-of-sight communications, but it can still only handle stationary users so is sometimes known as *fixed WiMAX*. IEEE 802.16e (2005), or *mobile WiMAX*, supports important aspects of mobility such as moving devices, handover and roaming.

### 1.4.4 Market history

The last two diagrams show the current state of the mobile cellular market. Figure 1.15 shows the geographical distribution of different cellular technologies in January 2008, with the area of each sector proportional to the number of subscribers. cdmaOne and cdma2000 are collected together, while the sectors labelled 'other' are mainly iDEN in the USA and Canada, D-AMPS in Latin America, and PDC in Asia and the Pacific.

Figure 1.16 shows the total numbers of subscribers to the different technologies since 2000, with historical data up to the beginning of 2008 and forecasts thereafter. The forecasts suggest that UMTS will gradually replace GSM, with its main areas of growth being Western Europe and East Asia. However, this process will be far from complete even by 2013.

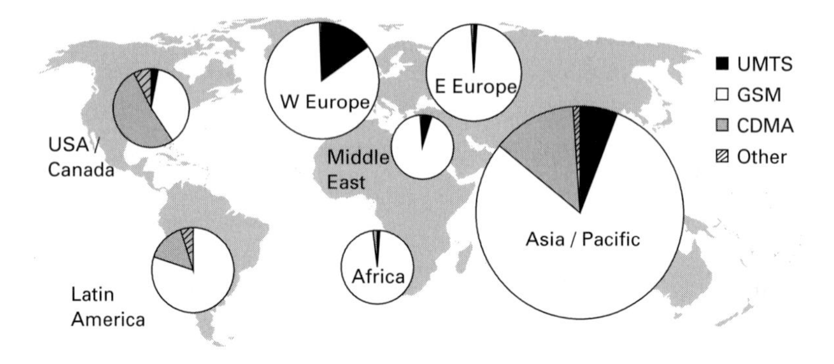


**Figure 1.15** Numbers of subscribers to different mobile communication technologies in 2008. (Data used with permission from Informa Telecoms & Media.)

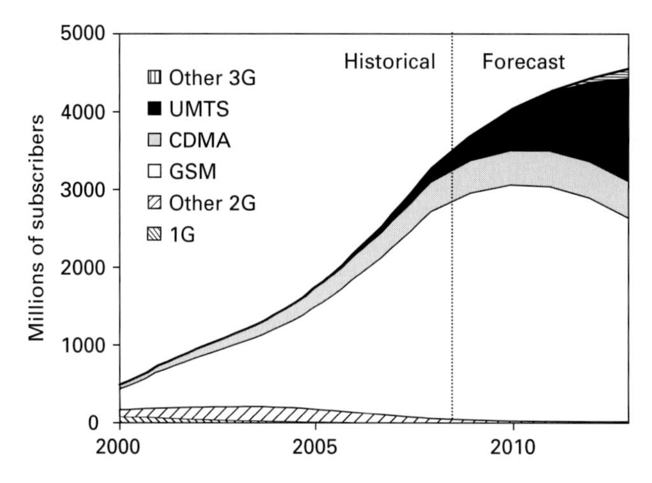


**Figure 1.16** Growth in the use of different mobile telecommunication technologies, with historical data from 2000 to 2008, and forecasts from 2008 to 2013. (Data used with permission from Informa Telecoms & Media.)

## References

1. J. Bannister, P. Mather & S. Coope, *Convergence Technologies for 3G Networks: IP, UMTS, EGPRS and ATM* (Wiley, 2003).
2. J. F. Kurose & K. W. Ross, *Computer Networking: A Top-Down Approach*, 4th edition (Addison Wesley, 2007).

3. W. Stallings, *Data and Computer Communications*, 8th edition (Prentice Hall, 2006).
4. A. S. Tanenbaum, *Computer Networks*, 4th edition (Prentice Hall, 2002).
5. J. Proakis & M. Salehi, *Digital Communications*, 5th edition (McGraw-Hill, 2007).
6. C. E. Shannon, A mathematical theory of communication. *Bell Systems Technical Journal*, **27** (1948), 379–423 and 623–656.
7. Informa UK Limited, *Mobile Content and Services*, 6th edition (2006).

# 2 Introduction to UMTS

This chapter serves as a system level introduction to UMTS. We begin by describing the 3rd Generation Partnership Project, which is the organisation that defines the architecture and operation of the system. We continue by examining the architecture of UMTS, the interfaces between the different hardware components, and the protocol stacks that they use. At the end of the chapter are two shorter sections that describe the data flows within the system and the allocation of frequency spectrum to third generation systems. By the end of this chapter, you should have an appreciation of how the system fits together, and be ready to take on the details that are covered later in the book.

## 2.1 The 3rd Generation Partnership Project

Most of the information in this book originates in the specifications that define the architecture and operation of UMTS. These specifications are written by an organisation called the *3rd Generation Partnership Project* (3GPP). In this section, we will describe how 3GPP is organised, and go on to discuss the specifications themselves.

### 2.1.1 Organisation of 3GPP

The 3rd Generation Partnership Project was formed in December 1998, to produce the technical specifications for UMTS. The formation of 3GPP came during the International Telecommunication Union's selection process for 3G telecommunication systems, and the first set of specifications was used as the member organisations' submission to the ITU. More recently, 3GPP has developed the UMTS specifications further and has expanded its role to handle the specifications for GSM, which had previously been produced by the *European Telecommunications Standards Institute* (ETSI).

Table 2.1 *3GPP organisational partners.*

| Organisation | Abbreviation | Country |
|---|---|---|
| Association of Radio Industries and Businesses | ARIB | Japan |
| Alliance for Telecommunications Industry Solutions | ATIS | USA |
| China Communications Standards Association | CCSA | China |
| European Telecommunications Standards Institute | ETSI | Europe |
| Telecommunications Technology Association | TTA | Korea |
| Telecommunication Technology Committee | TTC | Japan |

The project is structured as a collaboration between a number of telecommunication standards bodies that are known as organisational partners (Table 2.1). Each organisational partner has a membership comprising a large number of telecommunication companies, and the specifications are written by representatives from those member companies. As with other telecommunication standards, the member companies own a considerable amount of intellectual property that has been included in the UMTS specifications, and the resultant royalties can be a significant cost for manufacturers.

Internally, 3GPP is organised into a *project co-ordination group* (PCG) that handles the overall project management, and four *technical specification groups* (TSGs) that do the actual technical work. The individual TSGs (Table 2.2) meet either four or five times a year, to agree on updates to their respective specifications. Within each TSG are a number of different working groups, each of which concentrates on a particular part of the system.

There are many other telecommunication standards bodies, which produce specifications for the other systems described in Chapter 1, and for application software that is independent of the underlying

Table 2.2 *3GPP technical specification groups.*

| Technical specification group | Abbreviation |
|---|---|
| Services and system aspects | TSG SA |
| Core network and terminals | TSG CT |
| Radio access network | TSG RAN |
| GSM EDGE radio access network | TSG GERAN |

communication network. The only one we will mention is the *3rd Generation Partnership Project 2* (3GPP2), which is a parallel body to 3GPP that produces the specifications for cdma2000.

### 2.1.2 3GPP specifications

We now move on to the specifications, which are the actual documents that define the system. At a high level, the specifications are organised into *releases*, each of which is a version of the system with a particular set of features. 3GPP maintains the specifications for all the releases of UMTS in parallel. This allows it to add new features to the system as part of each new release, while making the occasional technical correction to the older, more stable releases that are used by manufacturers.

Each release is developed over a period of months or even years, but the most important event happens when the release is frozen. After it has been frozen, there are no more changes to a release's technical features, although some issues such as the details of the protocols and the conformance tests will usually lag behind. Technical corrections can of course continue for a long time after freezing.

The first release of UMTS was *release 99*, which was frozen in March 2000. This release specified a 3G telecommunication system based on the core network of GSM, but with a new air interface that used wideband code division multiple access (W-CDMA). Most of the technical features described in this book were introduced in release 99. The plan was then to have one release per year, using a numbering scheme of release 00, release 01 and so on. However, it was soon realised that this was too ambitious, so

Table 2.3 *Releases of the UMTS specifications, highlighting the date when the release was frozen and the most important new features to appear.*

| Release | Date frozen | New features |
|---|---|---|
| 99 | March 2000 | W-CDMA air interface |
| 4 | March 2001 | Bearer independent CS architecture |
| | | TD-SCDMA |
| 5 | June 2002 | HSDPA |
| | | IP multimedia subsystem |
| 6 | March 2005 | HSUPA |
| 7 | September 2007 | HSPA+ |
| 8 | | Long Term Evolution |

the numbering scheme was changed to uncouple it from the calendar year, and the next release became known as release 4. Using this scheme, release 99 is synonymous with release 3, while the numbers 1 and 2 are reserved for draft specifications.

Table 2.3 lists the different releases of UMTS, together with their freeze dates and their most important technical features. (At the time this book was written, release 8 was at an early stage of the specification process.) It is worth noting that equipment manufacturers and network operators do not have to implement all the features of a particular release. Instead, some features (such as the definitions of the signalling messages) are mandatory, while others (such as whether or not to implement high speed packet access) are optional.

Within each release, the different specifications are organised into series, each of which covers a different part of the system. Series 21 to 36 describe UMTS, including aspects of the system that are common with GSM, and these are listed in Table 2.4. Other series refer to features that are unique to GSM: series 00 to 13 were used up to release 99, and series 41 to 55 are for release 4 onwards.

Individual specifications have document numbers like (for example) TS 25.331 v 6.12.0. Here, TS stands for technical specification – there are also documents that do not actually define any part of the system, which are known as technical reports and denoted TR; 25 is the series

Table 2.4 *List of the 3GPP specification series that refer to UMTS.*

| Series | Description |
|---|---|
| 21 | Requirement specifications |
| 22 | Stage 1 service specifications |
| 23 | Stage 2 service specifications; network architecture |
| 24 | Air interface non-access stratum |
| 25 | Air interface access stratum; radio access network |
| 26 | Codecs |
| 27 | R interface |
| 28 | Tandem-free operation of speech codecs |
| 29 | Core network |
| 30 | 3GPP programme management |
| 31 | Cu interface |
| 32 | Network management and charging |
| 33 | Security procedures |
| 34 | UE test specifications |
| 35 | Cryptographic algorithms |
| 36 | Air interface Long Term Evolution |

number; 331 is the specification number within that series; 6 is the release number; 12 is the technical version number (which is incremented after technical changes to a specification); and 0 is the editorial version number (incremented after non-technical changes). This particular specification describes the *radio resource control* (RRC) protocol, which we will say a lot about in the course of the book.

There are several hundred specifications altogether, which can be downloaded from the 3GPP website, www.3gpp.org. They are rather heavy reading, however, as their only purpose is to define precisely how the system operates, without any attempt at context or explanation.

## 2.2 System architecture

In this section, we will discuss the architecture of UMTS, starting from a high level and moving down to the individual components. Most of the components have long-winded names, and are usually referred to by acronyms. It is worth taking the trouble to learn these acronyms, as they will make regular appearances in the later chapters of the book, and will be essential if the reader is to have a fluent conversation about UMTS.

### 2.2.1 High level architecture

Figure 2.1 shows the high level architecture of a UMTS public land mobile network. In keeping with the generic architecture that we introduced in Chapter 1, it has three main parts: the core network, the radio access network and the mobile.

The core network (CN) contains two *domains*. The circuit switched (CS) domain transports voice calls using circuit switched technology. It has interfaces to fixed line telephone systems that are known as *public switched telephone networks* (PSTNs), and to circuit switched domains that are run by other network operators. The packet switched (PS) domain transports data streams using packet switching. It communicates with data servers that are controlled by the network operator itself, with external *packet data networks* (PDNs) like the Internet, and with packet switched domains that are controlled by other network operators. The two domains were carried over from GSM and GPRS respectively, with only a few modifications.

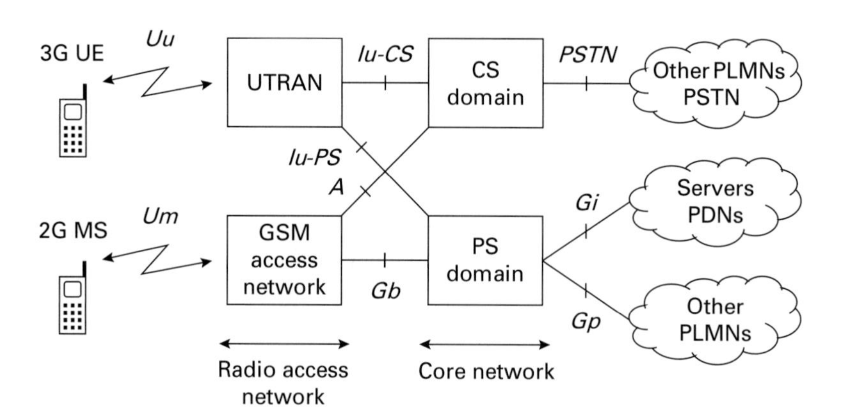


**Figure 2.1** High level architecture of UMTS.

The most important part of the radio access network is the *UMTS terrestrial radio access network* (UTRAN), which was introduced as part of release 99. However the system continues to support the GSM radio access network as well, to provide backwards compatibility with GSM. It is likely that other types of radio access network, such as satellite-based access, will eventually be introduced as well.

The UMTS mobile is known as the *user equipment* (UE). This is a change in terminology from GSM, where it was known as the *mobile station* (MS). Most UMTS mobiles are actually dual mode devices that support GSM as well: they communicate using 3G technology in regions of UMTS coverage, but revert to 2G in regions where UMTS base stations have not yet been deployed.

The figure also labels the interfaces between the different parts of the network. Most of these are for reference, but it is worth remembering the two types of Iu interface (Iu-CS and Iu-PS) that lie between the UTRAN and the core network, and the Uu interface between the UTRAN and the mobile. We will describe the different interfaces in more detail in Section 2.3, along with their associated protocol stacks.

The internal details of the core and radio access networks are rather complex, and have evolved somewhat since the release 99 specifications were written. We therefore describe the components of the release 99 architecture first, and then cover the changes that have taken place in releases 4 to 7. We will leave components that are just used by individual services until Chapter 5, and we will also delay any discussion of release 8 until Chapter 6.

### 2.2.2 Core network

Figure 2.2 shows the internal architecture of the release 99 core network, for both the circuit switched and packet switched domains. The notation for switches and databases is carried over from Figure 1.1, although this is only for guidance as some of the components are more flexible than these simple names imply. Dotted lines denote interfaces used only for signalling, while solid lines denote interfaces used for both traffic and signalling.

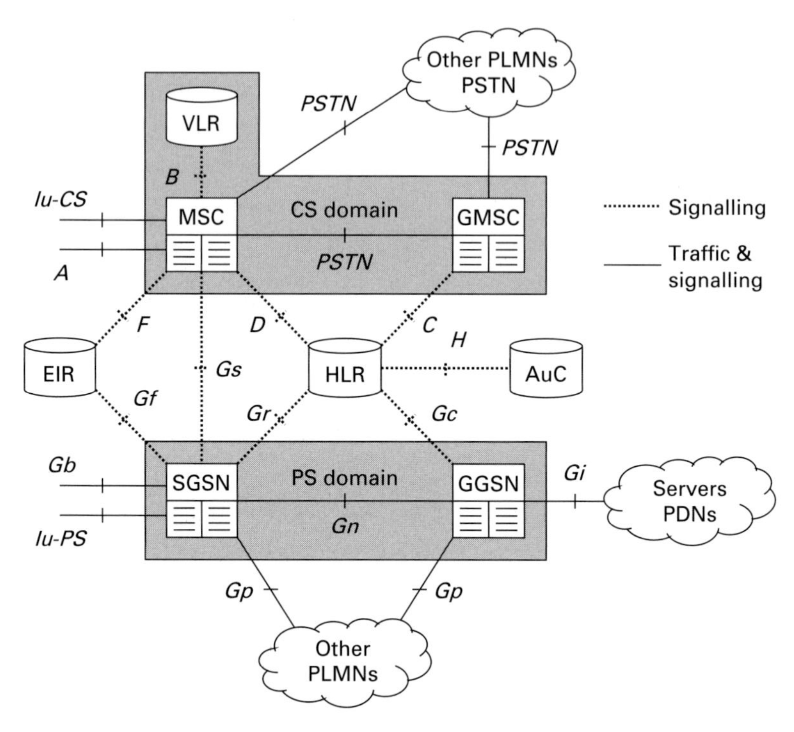


**Figure 2.2** Architecture of the core network in release 99. (Adapted from 3GPP TS 23.002.)

A few components are shared between the two domains. The most important of these is the *home location register* (HLR), which is the network operator's central database. The HLR contains information about the operator's subscribers such as their identities, their current locations and the services they have subscribed to. The *authentication centre* (AuC) contains security related information about the subscribers. Examples include secure keys that the network uses to confirm their identities and prevent unauthorised access. The *equipment identity register* (EIR) is an optional component: if implemented, it contains information such as a list of stolen mobiles.

The main component in the circuit switched domain is the *mobile switching centre* (MSC). A small network can just contain one MSC, but most networks have more than one, each of which looks after a particular geographical area known as an MSC area. The MSC acts as a switch for voice calls, and it also handles signalling communications with the mobiles that are in its MSC area. An MSC may be designated as a *gateway MSC* (GMSC), which acts as a point of entry into the network for incoming calls.

The *visitor location register* (VLR) looks after one or more geographical regions known as *location areas*. Each VLR contains a local copy of the HLR's information about the mobiles in its location areas, which minimises the communication needed between the two. The MSC and VLR are usually implemented as a single piece of hardware, so the interface between them does not actually exist as a physical entity. For this reason, we often won't bother to distinguish them in the chapters that follow, referring to them instead as a single MSC/VLR combination.

There are two components in the packet switched domain. The *serving GPRS support node* (SGSN) combines the functions of the MSC and the VLR by acting as a router for data transfers, keeping a local copy of information about the mobiles in its SGSN area, and handling all the signalling communications with those mobiles.

The *gateway GPRS support node* (GGSN) is rather different from the gateway MSC, however. It acts as an interface to data servers and to other networks for both incoming and outgoing data streams. It does not look after a geographical area in the same way that an SGSN does, although one piece of hardware can implement both sets of logical functions.

If a mobile is roaming, then the home network contains the HLR and AuC, and the visited network contains the MSC, VLR and SGSN. The gateways to external networks can be in either the home network or the visited one, depending on the circumstances.

It is worth remembering all the hardware components introduced here, as most of them will make a lot of appearances as the book goes on. The interface names are less important: we will refer to them when we describe the protocol stacks in Section 2.3, but hardly at all thereafter.

Before closing this section, it is useful to introduce a few numbers that are used to identify the network. The *mobile country code* (MCC) is a three-digit number that identifies the country that a network is in, while the *mobile network code* (MNC) is a two- or three-digit number that identifies a network operator in that country. In the UK, for example, the mobile country code is 234, while T-Mobile uses a mobile network code of 30. Together, the two numbers make up the *public land mobile network identity* (PLMN-ID).

### 2.2.3 Radio access network

The radio access network is shown in Figure 2.3. The most important part for us is the UMTS terrestrial radio access network (UTRAN), which has two components: the *Node B* and the *radio network controller* (RNC). The Iub interface connects a Node B to an RNC, while the Iur interface connects two RNCs. All the interfaces in the figure carry both traffic and signalling.

The *Node B* is a UMTS base station. It controls one or more cells (three in the figure), and transmits and receives radio signals to and from the mobiles that are in those cells; roughly speaking, it implements the physical layer of the air interface. The name 'Node B' does not stand for anything in particular: it was introduced as a temporary name, but the name stuck.

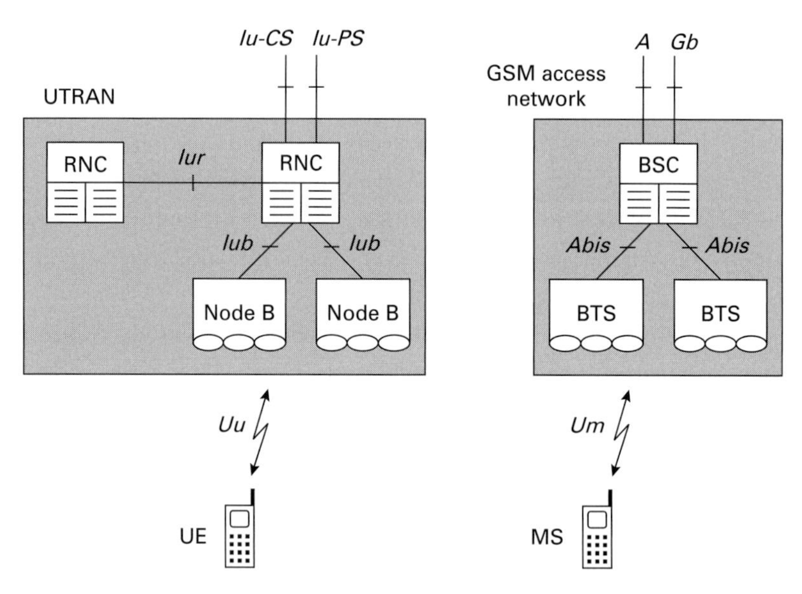


**Figure 2.3** Architecture of the radio access network in release 99. (Adapted from 3GPP TS 23.002.)

The radio network controller is an intermediate component between the Node B and the core network. The RNC has three main functions, and an individual RNC can acquire up to three different names depending on which of these functions it is implementing. First, each Node B is controlled by a particular RNC, which is known as its *controlling RNC* (CRNC). A controlling RNC distributes downlink traffic to the Node Bs that it controls, collects traffic from them on the uplink, and exchanges signalling messages with them. Second, each mobile is controlled by a particular RNC, which is known as its *serving RNC* (SRNC). A serving RNC exchanges signalling messages with the mobiles that it serves, and acts as their sole point of contact with the core network. It also implements the layer 2 communications between the mobile and the network, for example by handling any retransmissions that are required over the air interface.

In some situations, a serving RNC may not actually control the Node B that a mobile is communicating with. This situation is handled by introducing a third function (Figure 2.4a), that of a *drift RNC* (DRNC). A drift RNC uses the Iur interface to carry mobile specific traffic and signalling messages between the Node B and the serving RNC.

In other situations, a mobile can communicate with more than one cell at a time. This state is known as *soft handover*, and is illustrated in

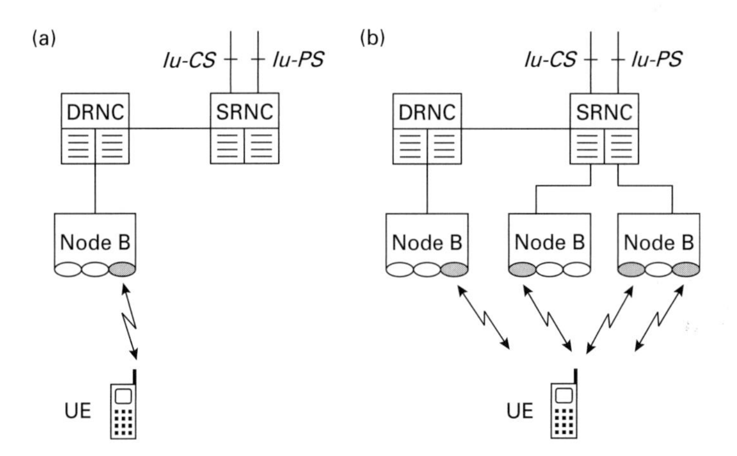


**Figure 2.4** Example geometries for the air interface. (a) Use of a drift RNC. (b) Soft handover.

Figure 2.4b. As shown in the figure, the cells can be controlled by the same Node B, or by different Node Bs, or even by different RNCs. If more than one RNC is involved, then one acts as the serving RNC, and the others act as drift RNCs. The cells used in soft handover are collectively known as the *active set*. We will have more to say about the implementation of soft handover in later chapters.

The system also supports the GSM radio access network, which is known as the *base station subsystem* (BSS). The main components in the BSS are the *base transceiver station* (BTS) and the *base station controller* (BSC). These are roughly analogous to the Node B and the RNC, but the division of functions between them is rather different, and release 99 BSCs have no equivalent of the Iur interface.

The internal structure of the UTRAN is important, and we will refer to it in several places as the book goes on. We will not have much to say about the GSM radio access network, however.

### 2.2.4 User equipment

Figure 2.5 shows the internal architecture of the user equipment (UE). There are two main components, the *mobile equipment* (ME) and the *universal integrated circuit card* (UICC). The ME is the mobile phone itself, while the UICC is a smart card that plugs into the mobile phone.

In a simple mobile phone, the ME is usually a single device, but in data terminals, its functions are often split in two: the *mobile termination* (MT) handles all the 3G communication functions, while the *terminal equipment* (TE) is the point where the data streams begin and

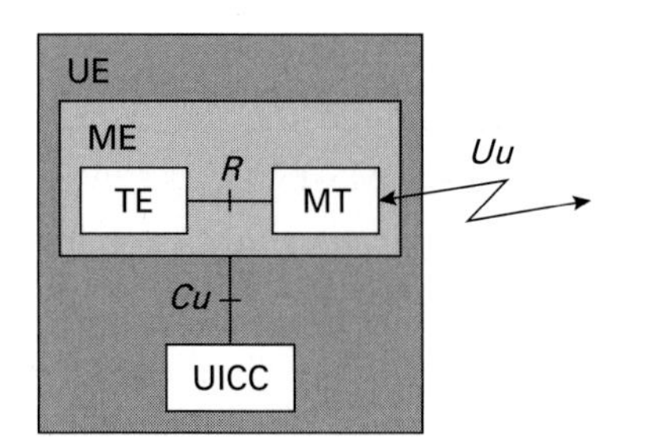


**Figure 2.5** Internal architecture of the mobile. (Adapted from 3GPP TS 27.001.)

end. The MT might be a plug-in UMTS card for a laptop, for example, while the TE might be the laptop itself.

The UICC evolved from the subscriber identity module (SIM), which was first introduced in GSM. The terminology has changed because UMTS makes a clear distinction between hardware and software: the UICC is the smart card hardware, while the *universal subscriber identity module* (USIM) is a software protocol that runs there. The UICC stores data that are associated with the subscriber and the network operator. Examples include the subscriber's identity, the services the subscriber can use, and the subscriber's personal phone book. However, the UICC is not just a data store. It also carries out calculations that are related to the network's security procedures, and it can run application software as well. The use of a UICC makes it easy for people to upgrade their mobile phones, by simply switching the UICC from one phone to another. If there is no UICC installed, the only thing the mobile can do is make emergency calls.

Mobiles have a wide range of capabilities, and can vary in parameters such as the highest data rate they can handle and the maximum number of simultaneous data streams they support. As a simple example, speech mobiles can only handle a low data rate, but for mobiles that support video the maximum data rate is much higher. Similarly, most UMTS mobiles can also communicate using GSM, but some are unable to do so.

In this book, we will normally refer to the user equipment as the mobile or the UE. Most of the functions we will describe are implemented in the MT, because this is the part that handles the 3G communication functions. However, we will only distinguish the UE's components when there is a good reason to do so, usually when discussing its internal operation.

Earlier, we introduced some numbers that describe individual networks, and we can now do the same thing for the mobile. The *mobile station integrated services digital network number* (MS-ISDN) is simply the user's phone number, and has the same format as phone numbers everywhere. The *international mobile equipment identity* (IMEI) is a unique identifier for the ME: it contains a type allocation code and a serial number. Similarly, the *international mobile subscriber identity*

(IMSI) identifies the subscriber and the UICC: it consists of the mobile country code and mobile network code, followed by a number that identifies the subscriber within the specified network. The IMEI and IMSI are only used for bookkeeping purposes: to avoid the risk of cloning, they are transmitted as rarely as possible and should never be released to the outside world.

### 2.2.5 Enhancements in later releases

The fixed network has had a number of architectural enhancements in releases 4 to 7. The details are less important than those of release 99, so we will only describe them briefly. To illustrate the main enhancements, Figure 2.6 shows the architecture of the core network at the end of

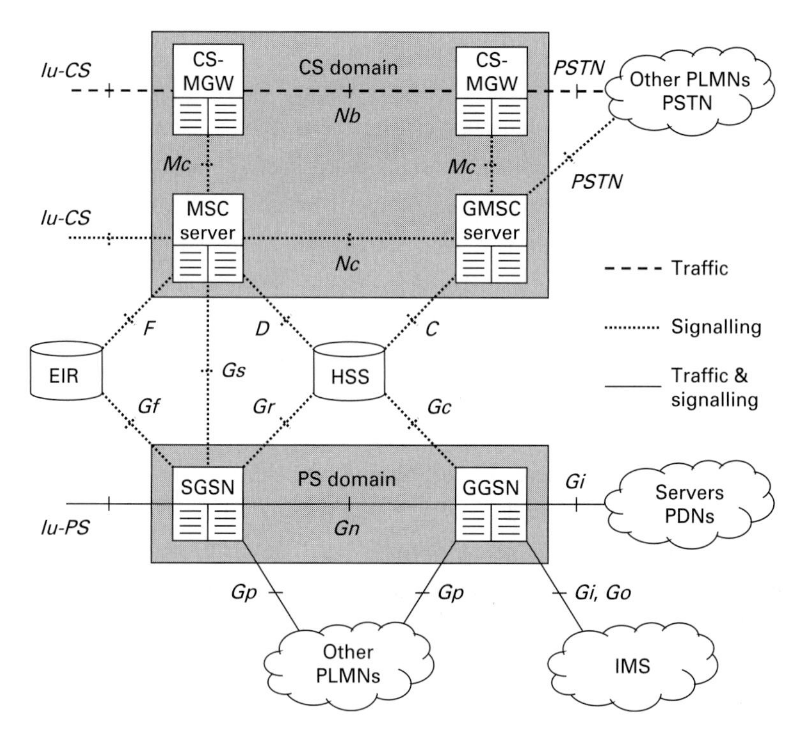


**Figure 2.6** Architecture of the core network in release 7. (Adapted from 3GPP TS 23.002.)

release 7. Dotted and solid lines are used in the same way as before, while interfaces shown with dashed lines are only used for traffic. The changes to the radio access network are relatively few.

The first enhancement was the *bearer independent circuit switched core network* in release 4. In this architecture, the mobile switching centre is split in two. The *circuit switched media gateway* (CS-MGW) handles the traffic functions of the MSC, but uses different transport protocols that we will see in the next section. It also includes a media conversion function, which allows it to communicate with networks that are using other types of transport protocol. The *MSC server* combines the signalling functions of the MSC with those of the VLR, and also controls the CS-MGW over a signalling interface that lies between them. A GMSC server is built in the same way.

The main network enhancement in release 5 is the *IP multimedia subsystem* (IMS). This is an extra network which interfaces with the packet switched domain, and which provides users with real time packet switched services that cannot be supplied using the packet switched domain alone. The *home subscriber server* (HSS) was also introduced in release 5, and combines the functions of the HLR and the AuC. The third release 5 enhancement (not shown in the figure) is an architectural feature known as *IuFlex*. In earlier releases, each radio network controller was connected to just one MSC and one SGSN. IuFlex introduces a more flexible architecture in which each RNC can be connected to multiple MSCs and multiple SGSNs.

The main release 6 enhancement is *wireless local area network* (WLAN) *interworking*. This allows users to access the network operator's packet switched services using a wireless LAN. The services are supplied either by the IMS, or by data servers that are controlled by the network operator and directly connected to a GGSN. The connection uses some extra core network components that are not shown in the figure, known as the *WLAN access gateway* (WAG) and *packet data gateway* (PDG).

There have also been improvements to the GSM radio access network, which is known as the *GSM EDGE radio access network* (GERAN) from release 4. The network has a new interface between base station controllers which is denoted Iur-g, and which includes some of the features of the Iur interface. A release 4 BSC can also communicate with the core network using the same Iu interface protocols that are used by UMTS, instead of the older protocols that were used by GSM.

## 2.3 Interfaces and protocols

A UMTS network contains a large number of interfaces, each of which has its own protocol stack. In this section, we will discuss the protocols that are used for signalling, transport and data manipulation, and show how those protocols are combined into stacks on the individual interfaces. Some of the protocols are more important than others, and the more important ones will be highlighted in the text. As in the case of the hardware components, it is worth learning the names of the important protocols, as they will make a lot of appearances later on in the book.

### 2.3.1 Introduction

The protocol stacks vary a lot from one interface to another, but they all follow the basic pattern shown in Figure 2.7. Each protocol stack has two main layers. The application layer creates and interprets the UMTS signalling messages and manipulates the data streams, while the transport layer just transfers them from one network component to another. The mapping of these layers onto the open systems interconnection (OSI) stack is not always clear but, roughly speaking, the application layer contains OSI layers 5 to 7, while the transport layer contains OSI layers 1 to 4.

A protocol stack also has up to three planes. Roughly speaking, the user plane carries information intended for the user such as voice or

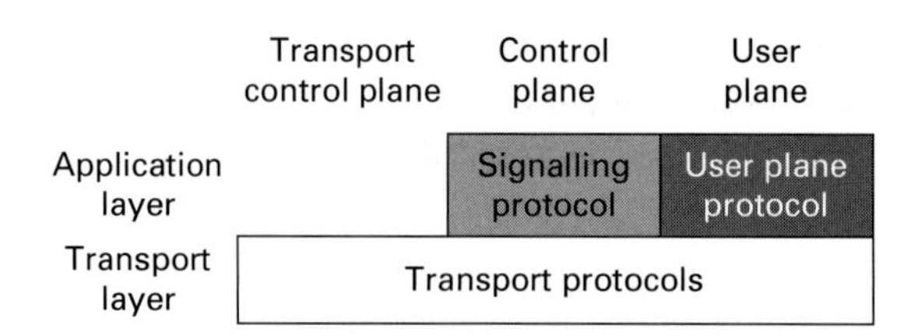


**Figure 2.7** Model of the protocol stacks in UMTS.

packet data, while the control plane carries signalling messages that are only of interest to the network. (There are a few cases where information swaps over between these planes: we will see some examples in due course.) If the data are being transported using ATM, then the transport control plane carries internal signalling messages that set up, modify and tear down any temporary virtual circuits that are required. These messages are only of interest to the transport layer, so the transport control plane does not extend any higher.

This lets us classify the protocols into three groups. In the application layer, the control plane contains signalling protocols that the network elements use to communicate with each other. The user plane protocols manipulate the data that the user is interested in, for example by compression and decompression. In the transport layer, most of the interfaces use the standard protocols that we introduced in Chapter 1. The big exception is the air interface, which uses protocols that are unique to UMTS. We will describe the protocols in these three groups in the sections that follow.

### 2.3.2 Signalling protocols

The first protocols to consider are the signalling protocols, in the application layer's control plane. We can illustrate how they operate by considering the *radio resource control* (RRC) protocol, which lies between the mobile and its serving radio network controller. Using this protocol, the SRNC sends signalling messages to the mobile to control how it behaves, and the mobile sends responses and information messages to the SRNC. The RRC specification contains precise definitions of the messages, their associated parameters, and the ways in which they are organised into signalling procedures. It's a lengthy document, the release 7 version coming to over 1200 pages.

Figure 2.8 shows an example signalling procedure, which the SRNC can use to find a mobile's capabilities. The procedure is shown as a *message sequence chart*, which indicates the messages that are exchanged between the mobile and the SRNC, together with the protocols that are responsible. It hides the low level details, such as the way in which the

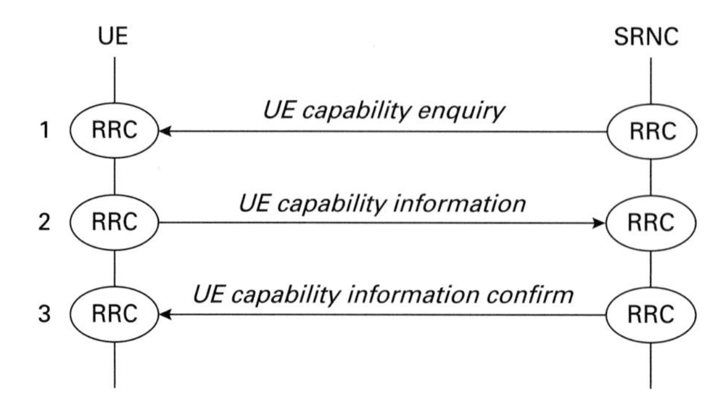


**Figure 2.8** Operation of the RRC protocol in the UE capability enquiry procedure.

messages are transported, and the way in which they pass through the Node B. (We will see some of those details later in the chapter.)

In step 1, the SRNC composes an RRC message known as *UE capability enquiry*, and sends it to the mobile. The mobile replies with a message called *UE capability information* (2), which includes several parameters that describe its capabilities. Examples include its maximum data rate, the number of simultaneous data streams it can handle, and whether or not it supports GSM. On receiving this information, the SRNC replies with a message called *UE capability information confirm* (3). It now knows how it should control the mobile, for example whether it can hand the mobile over from UMTS to GSM.

The procedures also define what happens in cases of abnormal behaviour. If, for example, a timer in the mobile expires before it receives the SRNC's confirmation message, then it is obliged to retransmit its capability information. We will not normally address cases of abnormal behaviour in this book; it will be enough to know that they exist.

UMTS has many different signalling protocols, which are listed in Table 2.5. They all behave in much the same way as the RRC protocol, in that they allow network elements to exchange signalling messages with each other.

It is worth highlighting the most important protocols in the table, as the operation of these protocols will be the main subject of Chapters 4 and 5. The *mobile application part* (MAP) handles signalling communications across most of the interfaces in the core network. For

Table 2.5 *List of signalling protocols in UMTS, grouped according to the part of the system where the protocol is used.*

| Location | Protocol | Description |
|---|---|---|
| CS domain | BICC | Bearer independent call control protocol |
| | ISUP | ISDN user part |
| | MEGACO | Media gateway control protocol |
| | TUP | Telephone user part |
| CS and PS domains | BSSAP+ | Base station subsystem application part plus |
| | MAP | Mobile application part |
| PS domain | GTP-C | GPRS tunnelling protocol control part |
| UTRAN | NBAP | Node B application part |
| | RANAP | Radio access network application part |
| | RNSAP | Radio network subsystem application part |
| Uu non-access stratum | CC | Call control |
| | GMM | GPRS mobility management |
| | MM | Mobility management |
| | SM | Session management |
| Uu access stratum | RRC | Radio resource control |
| UE | AT | Attention commands |
| | USIM | UMTS subscriber identity module |

example, when an incoming call arrives for a mobile at a gateway MSC, the GMSC sends a MAP message to the home location register in which it asks for the mobile's current location, so that it can forward the call to the correct MSC. The *Node B application part* (NBAP), *radio access network application part* (RANAP) and *radio network subsystem application part* (RNSAP) have similar roles within the radio access network, on the Iub, Iu and Iur interfaces respectively.

The air interface has two levels, the *non-access stratum* (NAS) and the *access stratum* (AS). Protocols in the non-access stratum exchange messages between the mobile and the core network. There are four of these. The *call control* (CC) protocol runs in the mobile and the circuit switched domain, and sets up, manages and tears down phone calls. Its companion in the packet switched domain is the *session management* (SM) protocol, which sets up, manages and tears down packet data transfers. The *mobility management* (MM) and *GPRS mobility management* (GMM) protocols handle bookkeeping messages that only affect the internal operation of the system, and are not related to any kind of data stream. The RRC protocol lies in the access stratum, and is used to exchange messages between the mobile and the radio access network.

### 2.3.3 Transport protocols

We now turn to the protocols in the transport layer, which are listed in Table 2.6. They fall into two distinct groups.

In the air interface's access stratum, information is transported using protocols that are unique to UMTS. The most important one is the air interface's physical layer. This carries out most of the processes that we described in Section 1.3, such as modulation, code division multiple access and error correction. We will spend much of Chapter 3 describing how this protocol works.

The physical layer is assisted by two layer 2 protocols. The *medium access control* (MAC) protocol controls the physical layer, for example by deciding how much data should be transmitted to or from a mobile at a particular time. The *radio link control* (RLC) protocol manages the data link between the mobile and the radio access network, by tasks such as retransmitting data packets if they arrive incorrectly. Roughly speaking, the physical layer is implemented in the mobile and Node B, while the MAC and RLC are implemented in the mobile and its serving RNC.

The fixed network uses standard transport protocols. In release 99, the circuit switched domain transmits voice calls using *pulse code modulation* (PCM), which is the transport mechanism used in digital fixed line telephone networks. In PCM, the analogue speech signal is digitised with 8 bit

Table 2.6 *List of transport protocols in UMTS, grouped according to the part of the system where the protocol is used.*

| Location | Protocol | Description |
|---|---|---|
| CS domain | PCM | Pulse code modulation |
| CS domain<br>PS domain | ALCAP | Access link control application protocol |
| UTRAN | ATM | Asynchronous transfer mode |
| | IP | Internet protocol |
| | MTP | Message transfer part |
| Uu access stratum | MAC | Medium access control |
| | PHY | Air interface physical layer |
| | RLC | Radio link control |

resolution at a sample rate of 8 kHz, to give a bit rate of 64 kbps. The resultant signal is converted to symbols, mixed with a carrier and multiplexed with other PCM signals, before transmission over the network. There is no other processing such as compression or error correction.

In other parts of the fixed network, data are transported using the protocols that we introduced in Section 1.2.3: asynchronous transfer mode (ATM), the Internet protocol (IP), and the message transfer part (MTP) of the SS7 protocol stack. As the system has evolved from one release to another, there has been a trend for IP to replace MTP and ATM, and we will see this trend reflected in the protocol stacks that follow.

The *access link control application protocol* (ALCAP) is only required if the data are being transported using ATM temporary virtual circuits. It is situated in the transport control plane, and sets up, manages and tears down the virtual circuits under command from higher layers.

There are several places in the fixed network where the actual protocol stacks are more complex than the ones we showed earlier, to handle issues like mixtures of different protocols. These issues are beyond the scope of this book, so in the protocol stacks that follow, we will simply describe the fixed network transport layers as 'ATM', 'IP over ATM' and so on.

### 2.3.4 User plane protocols

The user plane protocols manipulate the data that the user is interested in, in ways that are specific to UMTS. They also carry a small number of signalling messages, which control tasks that are closely related to their respective data streams such as timing synchronisation.

The best example is the *adaptive multi rate* (AMR) *codec*. In release 99, the circuit switched domain transports voice calls at a rate of 64 kbps. This is too fast for the air interface, because the signal-to-noise ratios are rather low there, so the maximum data rate is low. To deal with this problem, the AMR codec compresses the information on the path between the mobile and the MSC, to a rate between 4.75 and 12.2 kbps. This greatly increases the number of mobiles that a cell can support. From release 4 onwards, the information can be compressed all the way from one mobile to the other. If this is done, then the AMR codec is only implemented in the mobile, and the core network transports the information in compressed form using ATM or IP.

Table 2.7 lists the user plane protocols that are used by UMTS. Some of the protocols are only used by particular services, which are listed in the final column of the table. They are less important to us than the signalling or transport protocols, but a few of them will appear from time to time.

### 2.3.5 Circuit switched domain

We now have enough information to show the protocol stacks on each of the interfaces in UMTS. We will use the same shading scheme as in Figure 2.8, denoting the signalling protocols in grey and the user plane protocols with white letters on a dark background. We will also indicate the releases in which the later protocol options appear.

Figure 2.9 shows the protocol stacks in the circuit switched domain. On the internal signalling interfaces (Figure 2.9a), the network elements exchange signalling messages that are written using the mobile application part. These messages are transported using MTP in release 99, or using ATM or IP from release 4.

Table 2.7 *List of user plane protocols in UMTS, grouped according to the part of the system where the protocol is used.*

| Location | Protocol | Description | Service |
|---|---|---|---|
| CS domain | Nb UP | Nb user plane protocol | |
| PS domain | GTP-U | GPRS tunnelling protocol user part | GPRS |
| UTRAN | Iu UP | Iu user plane protocol | |
| | Iub UP | Iub user plane protocol | |
| | Iur UP | Iur user plane protocol | |
| Uu non-access stratum | AMR | Adaptive multi rate codec | Voice |
| | RLP | Radio link protocol | CS data |
| | SMS | Short message service protocol | SMS |
| Uu access stratum | BMC | Broadcast multicast control | CBS |
| | PDCP | Packet data convergence protocol | GPRS |
| UE | USAT | USIM application toolkit | |

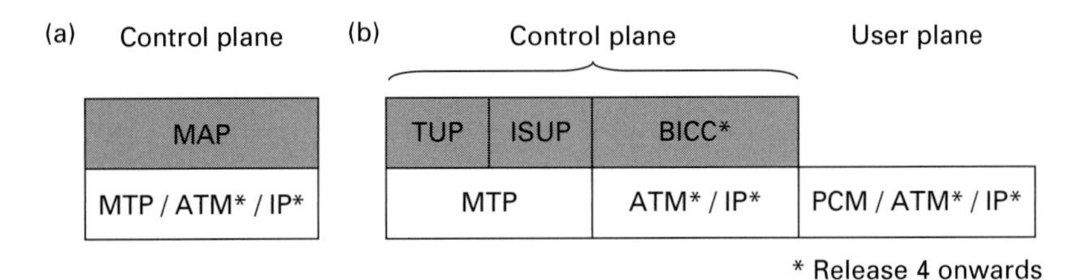


**Figure 2.9** Protocol stacks used by the circuit switched domain and the shared components of the core network. (a) Internal signalling interfaces (A to G, Gc, Gf and Gr). (b) PSTN interface to an external network.

The PSTN interface (Figure 2.9b) lies between the network and the outside world. In release 99, signalling messages are written using the ISDN user part or telephone user part, and are transported using MTP. From release 4, the *bearer independent call control* (BICC) protocol has the same functions as TUP or ISUP, but can use any underlying transport layer. Speech signals in release 99 are transported using pulse code modulation; from release 4, they can also be transported in compressed form using ATM permanent virtual circuits or using IP.

### 2.3.6 Bearer independent circuit switched domain

Figure 2.10 shows the new protocol stacks that were introduced in release 4, to control the bearer independent circuit switched core network. The Nb interface (Figure 2.10a) is a user plane interface between two media gateways, while Nc (Figure 2.10b) is a signalling interface between two MSC servers. Mc (Figure 2.10c) is the signalling interface by which an MSC server controls a media gateway, by means of the *media gateway control protocol* (MEGACO). These interfaces are less important to us than most of the others, and we will not discuss them further.

### 2.3.7 Packet switched domain

Figure 2.11 shows the protocol stacks in the packet switched domain. Between an SGSN and a GGSN (Figure 2.11a), the *GPRS tunnelling protocol user part* (GTP-U) routes data packets from one network element to another, using a mechanism known as tunnelling that will be covered in

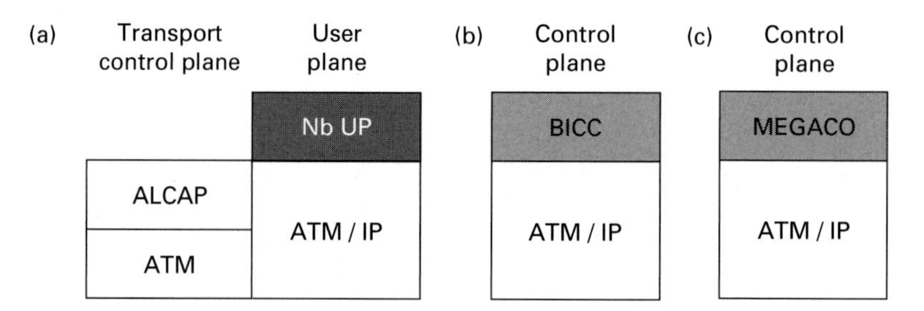


**Figure 2.10** Protocol stacks used by the bearer independent circuit switched domain from release 4 (a) Nb interface between two CS-MGWs. (b) Nc interface between two MSC servers. (c) Mc interface between an MSC server and a CS-MGW.

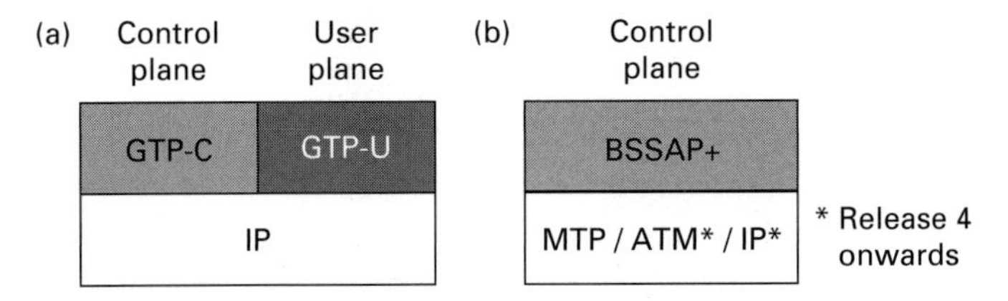


**Figure 2.11** Protocol stacks used by the packet switched domain. (a) Gn and Gp interfaces, between an SGSN and a GGSN. (b) Gs interface between an SGSN and an MSC.

Chapter 5. Signalling messages are handled by the *GPRS tunnelling protocol control part* (GTP-C), which has procedures for setting up, managing and tearing down tunnels, plus other functions that are similar to those of MAP.

The Gs interface (Figure 2.11b) transfers signalling messages between the circuit switched and packet switched domains, to support mobiles that are communicating with both. The signalling protocol is a modified version of the one used between the core network and the GSM radio access network, and is known as the *BSS application part plus* (BSSAP+).

### 2.3.8 Radio access network

The protocol stacks in the radio access network are shown in Figure 2.12. In these protocol stacks, the RANAP, RNSAP and NBAP protocols carry signalling messages between the different network elements. The user plane protocols carry data, plus a few closely related signalling messages that handle tasks such as timing synchronisation.

Depending on the particular interface being considered, the transport layer can use some or all of ATM, IP over ATM or (from release 5 onwards) IP alone. When ATM is used, the data are usually routed using temporary virtual circuits, which are managed by the access link control application protocol. On the Iu-PS interface, however, the data are always routed using IP, by means of the same tunnelling process that we noted above. On that interface, any ATM virtual circuits are permanent ones, so ALCAP is not required.

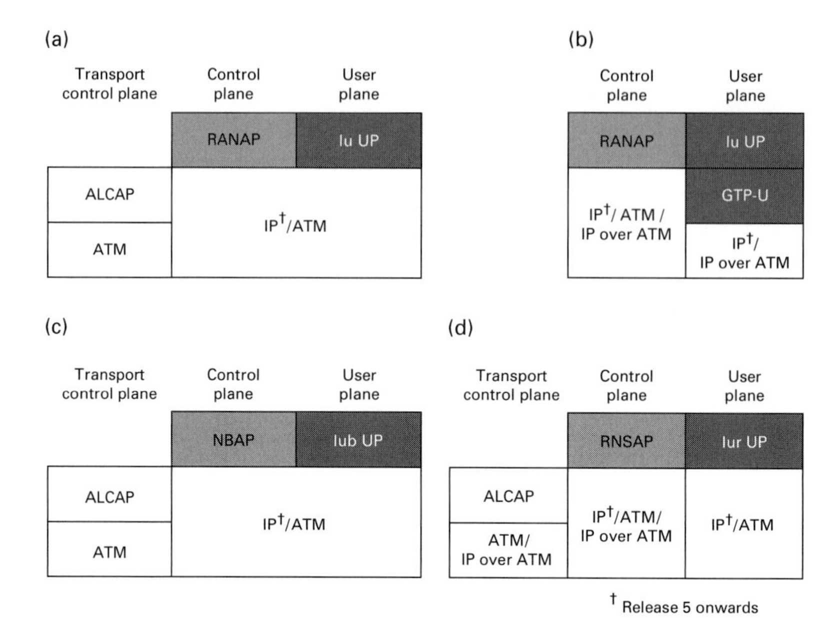


**Figure 2.12** Protocol stacks used by the radio access network. (a) Iu-CS interface between an RNC and the CS domain. (b) Iu-PS interface between an RNC and the PS domain. (c) Iub interface between an RNC and a Node B. (d) Iur interface between two RNCs. (Adapted from 3GPP TS 25.410, 25.420 and 25.430.)

## 2.3.9 Air interface

The interface between the mobile and the network is more complex than the others. It has two levels, known as strata, which are shown in Figure 2.13. The non-access stratum (NAS) protocols exchange data and signalling messages between the mobile and the core network. Examples include the call control and session management protocols that we saw in Section 2.3.2, and the AMR codec from Section 2.3.4.

The information is transported using the protocols in the two access strata below. The Uu access stratum (AS) contains the protocols that exchange data and signalling messages between the mobile and the radio access network, such as the RRC protocol that we saw earlier. Similarly, the Iu access stratum transfers information between the radio access

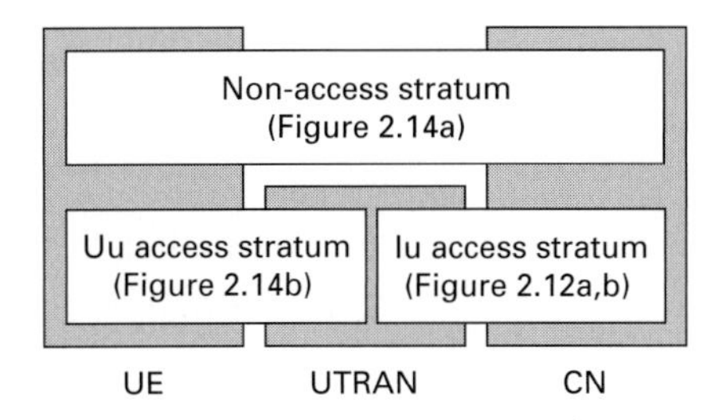


**Figure 2.13** Model of the air interface protocol stack. (Adapted from 3GPP TS 25.401.)

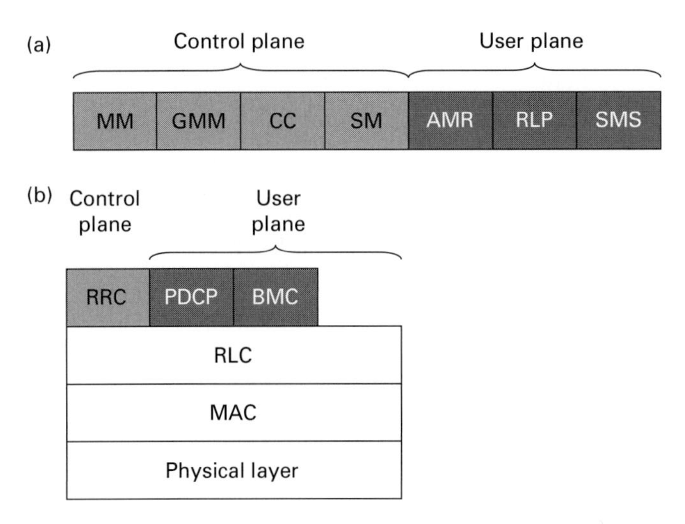


**Figure 2.14** Protocol stacks used by the air interface. (a) Non-access stratum between the mobile and the core network. (b) Uu interface access stratum between the mobile and the radio access network.

network and the core network. We have already seen the protocol stacks that it uses, in Figures 2.12a and 2.12b.

Figure 2.14a shows the protocol stack for the non-access stratum. We have already introduced most of the protocols here, notably the four signalling protocols (mobility management, GPRS mobility management, call control and session management), and the adaptive multi rate codec. The other two protocols are less important. The *radio link protocol* (RLP) handles circuit switched data streams such as fax transmissions, while the short message service (SMS) protocols transfer text messages between the mobile and the core network.

Figure 2.14b shows the protocol stack for the Uu access stratum. We have already seen the radio resource control protocol and the three transport protocols (radio link control, medium access control, and the air interface's physical layer). As before, the others are less important: the *packet data convergence protocol* (PDCP) compresses the headers of IP packets, and the *broadcast multicast control* (BMC) protocol handles a service known as the cell broadcast service.

### 2.3.10 User equipment

Inside the UE, the Cu interface (Figure 2.15a) lies between the mobile equipment and the universal integrated circuit card. The *universal subscriber identity module* (USIM) is best understood as a signalling protocol, while the *USIM application toolkit* (USAT) transfers data to and from applications that are running on the UICC. The transport protocols follow specifications for generic smart card platforms that are produced by the European Telecommunications Standards Institute (ETSI).

The R interface (Figure 2.15b) is used if the ME is split into the mobile termination and the terminal equipment. Most of this interface lies outside the scope of UMTS; instead, it uses a variety of open standards such as the universal serial bus (USB) and Bluetooth. However the specifications define a few issues, notably a set of *attention* (AT) commands that exchange signalling messages between the two devices.

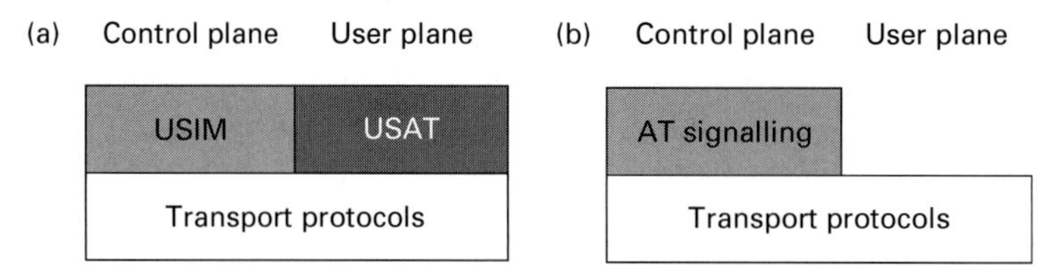


**Figure 2.15** Protocol stacks used inside the mobile. (a) Cu interface between the ME and the UICC. (b) R interface between the MT and the TE.

## 2.4 UMTS data streams

At this point, the reader is probably wondering how these protocol stacks fit together. There are many different answers, the choice depending on issues like the type of information being transferred and the configuration of the air interface. We will show a couple of examples later in this section.

First, however, we need to introduce two types of data stream that have special names in UMTS: bearers and channels. By analogy with the discussion of protocols in Section 1.2.2, these data streams can be viewed either as exchanges of information between different parts of the system, or as exchanges between the different layers of a single protocol stack. The first viewpoint turns out to be more useful for the bearers, while the second is more useful for the channels.

### 2.4.1 Bearers

A *bearer* is a data stream that spans some part of the system and has a specific *quality of service* (QoS). Figure 2.16 shows the most important bearers in UMTS.

When the mobile and the network agree to set up a data stream, the system first implements it using a *UMTS bearer*. This carries information

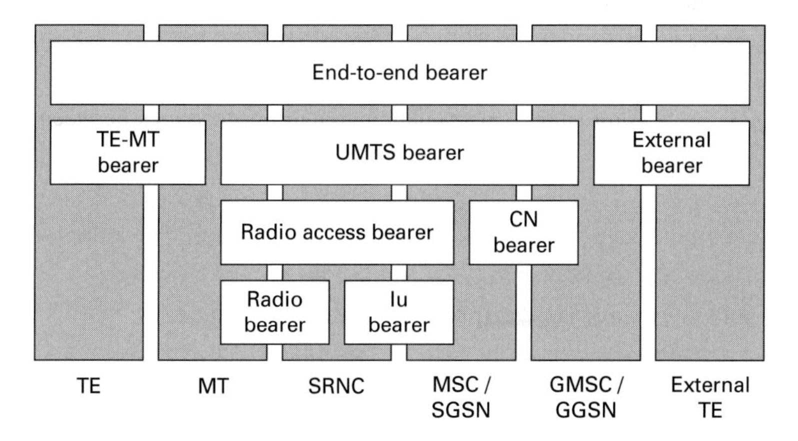


**Figure 2.16** Bearers used in UMTS. (Adapted from 3GPP TS 23.107.)

Table 2.8 *Signalling radio bearers in UMTS.*

| Bearer | Purpose |
|---|---|
| SRB 0 | Access stratum signalling for a UE in RRC Idle mode |
| SRB 1 | Access stratum signalling using RLC unacknowledged mode |
| SRB 2 | Access stratum signalling using RLC acknowledged mode |
| SRB 3 | High priority non-access stratum signalling |
| SRB 4 | Low priority non-access stratum signalling |

such as voice or packet data between the mobile termination and the far end of the core network (an MSC, GMSC or GGSN). If the MT and TE are implemented as two different devices, then another bearer transports information between them. However, this bearer lies outside the scope of UMTS, so we will not consider it further. The same applies to the bearer that lies beyond the far end of the core network.

The UMTS bearer is associated with a number of quality of service parameters. These describe the service that the user expects to receive, using parameters such as the required data rate, error rate and delay. The system cannot supply this quality of service right away, because the UMTS bearer spans different interfaces that use different transport protocols. It therefore breaks the UMTS bearer down into bearers that have a smaller scope. A *CN bearer* handles the path over the core network, while a *radio access bearer* (RAB) handles the path between the mobile and its first point of contact there. In turn, the radio access bearer is broken down into an *Iu bearer* between the core network and the SRNC, and a *radio bearer* between the SRNC and the mobile.

Each bearer is then implemented using the transport protocols that are appropriate for the corresponding interface, which provide the user with the quality of service expected. On the air interface, for example, the radio bearer is implemented using the RLC, MAC and physical layer protocols, which we will be describing in Chapter 3.

Five special radio bearers carry signalling messages between the mobile and its serving RNC. They are known as *signalling radio bearers* (SRBs), and are listed in Table 2.8. Each of them is implemented in a particular way that is appropriate for a particular type of message. SRB 0 is used to

Table 2.9 *Logical channels in UMTS release 99.*

| Channel | Name | Direction | Information carried |
|---|---|---|---|
| BCCH | Broadcast control channel | DL | System information messages |
| PCCH | Paging control channel | DL | Paging messages |
| CCCH | Common control channel | UL, DL | Signalling to/from a UE in RRC Idle mode |
| DCCH | Dedicated control channel | UL, DL | Signalling to/from a UE in RRC Connected mode |
| CTCH | Common traffic channel | DL | Broadcast/multicast traffic |
| DTCH | Dedicated traffic channel | UL, DL | Traffic to/from a single UE |

set up signalling communications between the mobile and the network; the other signalling radio bearers handle all subsequent communications. SRBs 1 and 2 carry RRC messages between the mobile and its serving RNC, the main difference between them being in the configuration of the RLC protocol. SRBs 3 and 4 are used to forward non-access stratum messages that begin or end in the core network. SRB 4 is optional, but if it is implemented, then SRB 3 is used for high priority messages, and SRB 4 is used for low priority ones.

### 2.4.2 Channels

In the air interface's access stratum (Figure 2.14b), the data flows between the different protocols are known as *channels*. There are many different types of channel, which are distinguished by the information they carry and the ways in which they are used. They can be collected into three different classes, which are used at three different levels of the protocol stack.

*Logical channels* (Table 2.9) lie between the radio link control and medium access control protocols. There are two ways to distinguish them.

Table 2.10 *Transport channels in UMTS release 99. (The list excludes a few channels that were deleted in later releases.)*

| Channel | Name | Direction | Information carried |
|---|---|---|---|
| BCH | Broadcast channel | DL | BCCH |
| PCH | Paging channel | DL | PCCH |
| FACH | Forward access channel | DL | Traffic and signalling on a common CDMA code |
| RACH | Random access channel | UL | Traffic and signalling on a common CDMA code |
| DCH | Dedicated channel | UL, DL | Traffic and signalling on a dedicated CDMA code |

The *dedicated traffic channel* (DTCH) and *dedicated control channel* (DCCH) are *dedicated logical channels*, and always carry information for a single mobile; the others are known as *common logical channels*, and can carry information for more than one mobile. *Logical traffic channels* carry user plane information such as voice or packet data, while *logical control channels* carry signalling messages in the control plane. For our purposes, the logical channels are the least important of the three classes, and we will not refer to them much.

*Transport channels* (Table 2.10) lie between the MAC and the physical layer. There is nearly a one-to-one mapping between the transport and physical channels, and the two classes can often be considered together. We will look at the information flows through the transport channels in a moment.

*Physical channels* (Table 2.11) lie below the air interface's physical layer. Roughly speaking, each physical channel is an allocation of a CDMA code for a particular purpose. There are two ways to distinguish them, but the distinctions are subtly different from the ones we saw earlier. The *dedicated physical data channel* (DPDCH) and *dedicated physical control channel* (DPCCH) are *dedicated physical channels*, in which the

Table 2.11 *Physical channels in UMTS release 99. (The list excludes a few channels that were deleted in later releases.)*

| Channel | Name | Direction | Information carried |
|---|---|---|---|
| PCCPCH | Primary common control physical channel | DL | BCH |
| SCCPCH | Secondary common control physical channel | DL | PCH and FACH |
| PRACH | Physical random access channel | UL | RACH |
| DPDCH | Dedicated physical data channel | UL, DL | DCH |
| PICH | Paging indicator channel | DL | Control signals to support the PCH |
| AICH | Acquisition indicator channel | DL | Control signals to support the RACH |
| DPCCH | Dedicated physical control channel | UL, DL | Control signals to support the DCH |
| SCH | Synchronisation channel | DL | Control signals to support acquisition |
| CPICH | Common pilot channel | DL | Pilot signal to assist the UE receiver |

CDMA code is assigned to a single mobile; in the others, known as *common physical channels*, the CDMA code can be used by more than one mobile. *Physical traffic channels* (the first four channels listed) map onto higher level channels, and carry both user plane data and control plane signalling messages. *Physical control channels* help the operation of the physical layer: they are composed in the physical layer of the transmitter and interpreted by the physical layer of the receiver, and are completely invisible to higher layers.

There are too many acronyms to remember easily, but there are really only five types of information flow, which correspond to the five transport channels. They are shown in Figure 2.17, in rough order of their

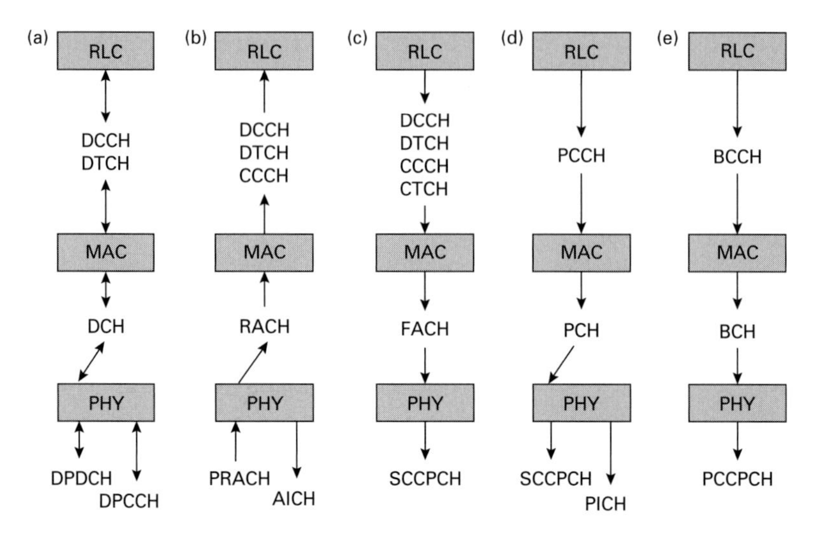


**Figure 2.17** Mapping of the logical, transport and physical channels for five information flows. (a) Use of a dedicated physical channel. (b) Use of a common physical channel in the uplink. (c) Use of a common physical channel in the downlink. (d) Paging messages. (e) System information broadcasts.

importance to us. The figure is drawn from the network's point of view, so that up and down arrows refer to the uplink and downlink respectively.

Figure 2.17a shows how information is sent using a dedicated physical channel, whose CDMA code is assigned to a single mobile. The *dedicated channel* (DCH) and the dedicated physical data channel carry data and signalling messages in both the uplink and downlink. They are assisted by the dedicated physical control channel, which sends physical layer control signals from the transmitter to the receiver. This is the usual scenario during a phone call; it is also the only scenario in which soft handover is supported. These three channels are the most important ones for this book, and we will examine their operation in some detail in Chapter 3.

Information can also be sent using a common physical channel, whose CDMA code can be used by more than one mobile. On the uplink (Figure 2.17b), the *random access channel* (RACH) and *physical random access channel* (PRACH) carry data and signalling messages from the mobile to the network. They are assisted by the *acquisition indicator channel* (AICH), which sends physical layer control signals in the opposite direction. On the downlink (Figure 2.17c), the information is sent on the *forward access channel* (FACH) and the *secondary common control physical channel* (SCCPCH). We will examine the operation of these channels in Chapter 3 as well.

Figure 2.17d is used for paging messages, which alert mobiles to incoming calls, while Figure 2.17e is used for system information broadcasts, which tell the mobiles in a cell about how the cell is configured. We will examine these two processes in Chapter 4.

Figure 2.17 left out two channels that only appear in the physical layer: the *common pilot channel* (CPICH) and the *synchronisation channel* (SCH). These are downlink-only channels which help the operation of the mobile's physical layer, and which will make an appearance in Chapters 3 and 4. There are also some more channels not listed here. First, release 99 defined some other channels that were later removed from the specifications because nobody implemented them, and we will not consider those channels at all. Second, more channels have been introduced in later releases of UMTS, particularly to support high speed packet access. We will describe those channels in Chapter 3.

### 2.4.3 Example signalling flows

We can now examine some of the information flows over the air interface, to show how the protocol stacks, bearers and channels fit together. There are many different scenarios, but a couple of examples will illustrate how it all works.

Figure 2.18 shows the protocols that are used to exchange signalling messages between the mobile and its serving radio network controller, for a scenario where the mobile is just communicating with one Node B and one RNC. These protocols might be used for the capability enquiry procedure that we saw in Figure 2.8.

First let's look at the downlink. The SRNC composes its message using the RRC protocol, and sends it on signalling radio bearer 1 or 2. It then processes the message using the RLC and MAC, and sends it to the Node B using the Iub user plane protocol. (This is one example in

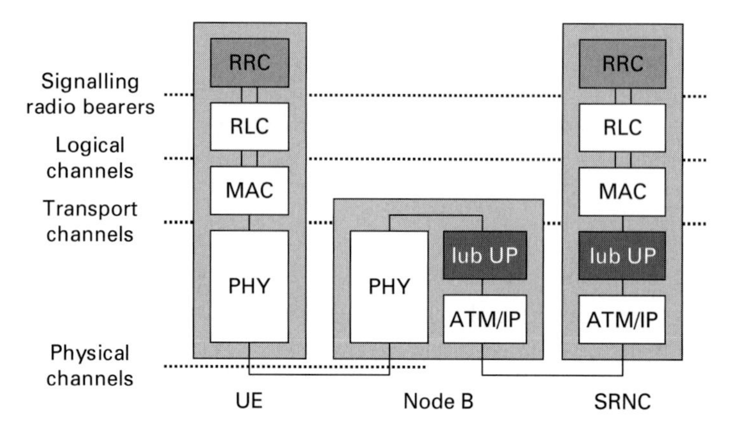


**Figure 2.18** Signalling flows between a mobile and an RNC, for a case where the mobile is communicating with one Node B and one RNC.

which a control plane message crosses into the user plane for part of its journey. The Iub control plane has a different purpose, namely signalling communications between a Node B and its controlling RNC using the Node B application part.) The Node B transmits the message using the DPDCH or SCCPCH, depending on whether it is using a dedicated or a common CDMA code, and the mobile receives and processes it. On the uplink, the flows are reversed, and the message is transmitted using the DPDCH or PRACH.

Figure 2.19 is a more complex scenario, which introduces three new features. First, the mobile is now in soft handover with cells that are controlled by two Node Bs. Soft handover is only supported when using dedicated channels, so on the air interface, the information is sent and received on the DPDCH. The receivers need some special mechanisms to combine the two signals, and we will describe those in Chapter 3. Second, one of the Node Bs is controlled by a different RNC, which is labelled as a drift RNC. Information is sent between the serving RNC and the drift RNC over the Iur interface, using the Iur user plane protocol.

Third, the figure shows the protocols that are used to exchange non-access stratum signalling messages between the mobile and the core network. These messages are transported across the Uu and Iu interfaces using the RRC and RANAP protocols, by embedding them into

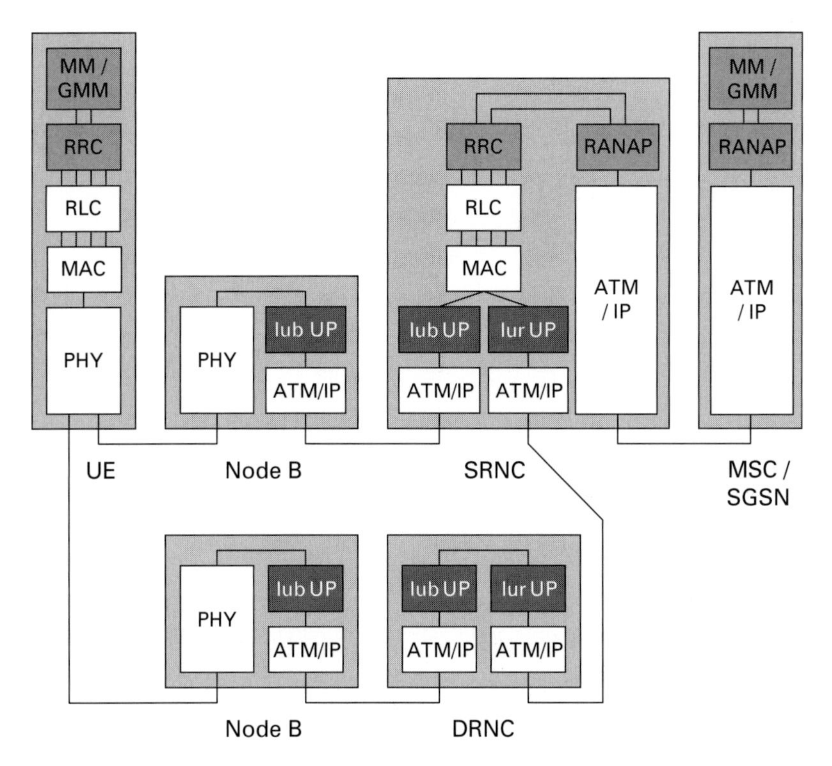


**Figure 2.19** Signalling flows between a mobile and the core network, for a case where the mobile is in soft handover, and is communicating with two Node Bs and two RNCs.

special access stratum messages that are known as *direct transfers*. On the Uu interface, the direct transfers are transported using signalling radio bearer 3 or 4.

Figure 2.20a shows an example. In this example, the circuit switched domain wants to confirm the subscriber's identity, perhaps because of a database failure in the core network. To do this, it sends the mobile a mobility management message called an *identity request* (1). In the message it specifies the type of identity that it requires, such as the international mobile subscriber identity (IMSI). The mobile replies with an MM message called *identity response* (2), which contains the identity requested.

The access stratum transports the messages as shown in Figure 2.20b. On the downlink, the core network's request is embedded into a

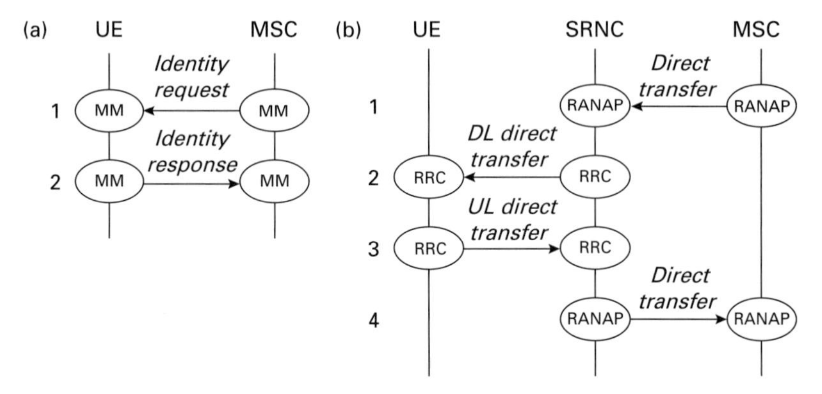


**Figure 2.20** Operation of the non-access stratum signalling protocols. (a) MM identification procedure. (b) Implementation of the procedure using RRC and RANAP direct transfers.

RANAP *direct transfer* (1), which is transported like any other RANAP message. The SRNC extracts the core network's request and embeds it into an RRC *downlink direct transfer* (2), which it sends to the mobile. Similarly, the mobile's response is embedded into an RRC *uplink direct transfer* (3), and another RANAP direct transfer (4). The use of different message names by the two protocols is unimportant; it is just an artefact of the specifications. We will see many non-access stratum signalling messages in Chapters 4 and 5, and they will all be transported in this way.

## 2.5 Frequency allocation

In this final section, we will look at some remaining details of the air interface: the modes in which the air interface can operate, and the way in which carrier frequencies are allocated to UMTS.

### 2.5.1 FDD and TDD modes

On the air interface, the physical layer transmits and receives radio signals using two distinct modes: *frequency division duplex* (FDD) and *time division duplex* (TDD). These are two techniques for distinguishing the mobiles' transmissions from those of the base stations, to ensure that they do not interfere with each other. FDD uses a paired spectral allocation, in which the base stations and mobiles transmit continuously but on two different frequencies. By contrast, TDD uses an unpaired spectrum, in which the base stations and mobiles transmit on the same frequency but at different times.

Both modes have advantages and disadvantages. FDD networks are easier to implement: they do not require accurate time synchronisation, and the use of two frequencies makes them less prone to interference. On the other hand, TDD uses the air interface more efficiently: if users are downloading more data than they are uploading, then the network can allocate more transmit time to the base stations than to the mobiles.

To date, FDD has proved much more popular than TDD, to the extent that there are very few implementations of the release 99 TDD specifications. Because of this, we will only cover FDD mode in this book. There is one exception, however: the Chinese TD-SCDMA system is actually an option within UMTS TDD mode that entered the specifications as part of release 4.

### 2.5.2 Worldwide frequency allocations

In UMTS, the individual carrier frequencies are 5 MHz apart, but they can lie in a number of different bands that are defined in the 3GPP specifications. Figure 2.21 shows the most important frequency bands in FDD mode.

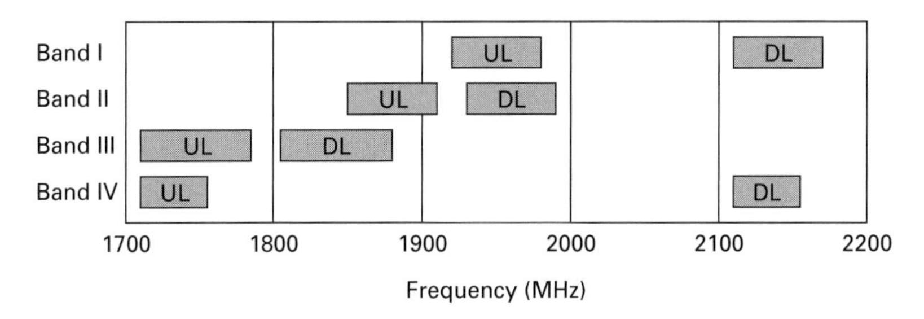


**Figure 2.21** Illustration of the most important frequency bands used by UMTS FDD mode.

Most of the world uses band I, which spans 1920–1980 MHz for the uplink and 2110–2170 MHz for the downlink, and provides 12 pairs of carrier frequencies. In North and South America, however, band I is unsuitable because it overlaps with existing allocations. In this region, UMTS is currently transmitted in band II, which actually has the same frequency as one of the 2G bands there: 1850–1910 MHz for the uplink, and 1930–1990 MHz for the downlink. Unfortunately, UMTS requires a much higher bandwidth than any second generation system (or even than cdma2000), so it is hard for network operators there to get licences for UMTS transmission. This is one of the reasons behind the slow take-up of UMTS in the Americas.

The specifications also support transmission in other frequency bands, which have been designed for the future expansion of UMTS. For example, band III is currently used for GSM 1800 transmissions in Europe and East Asia, while band IV is being made available for 3G-only transmissions in North and South America. Release 7 supports a total of ten bands for FDD, and four bands for TDD.

### 2.5.3 Allocations to network operators in the UK

The frequency bands shown above are awarded to individual network operators, usually on a country-by-country basis. To illustrate this, Figure 2.22 shows the current frequency allocations in the UK.

The licences were allocated by a government auction, which was designed by game theorists to extract as much money as possible from the network operators, and accidentally timed close to the peak of the dotcom boom in 2000. The result was that the operators paid a total licence fee of over £22 billion, or nearly £400 for each person in the country. A similar process happened in Germany, while in other countries the licences were awarded later and far more cheaply.

As shown in the figure, the UK auction included both FDD and TDD frequencies. However, the operators paid very much less for the TDD bands than they did for the FDD bands, and they have scarcely used them.

Note that an operator could run a UMTS network using just one pair of FDD frequencies, with the individual base stations and mobiles

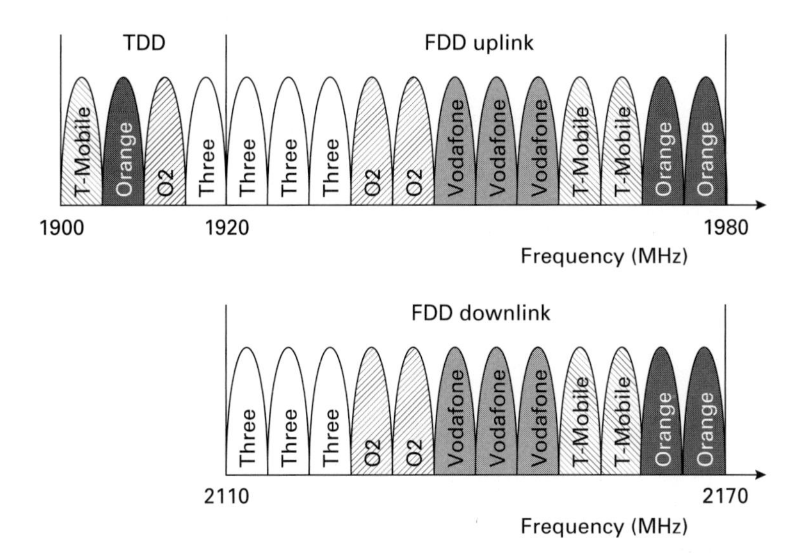


**Figure 2.22** Allocation of UMTS carrier frequencies to network operators in the UK. (Reproduced with permission from Imagicom Limited.)

distinguished by the use of CDMA codes. So how do they use multiple carrier frequencies? We will see in Chapter 3 that there is some interference between transmitters on the same frequency, which ultimately sets a limit on the capacity of the system. It turns out that this interference is particularly noticeable between cells of different sizes, so network operators tend to run macrocells on one FDD pair and microcells on another. Other carrier frequencies can be used to increase the capacity of the macro- or microcells, or for a third layer of picocells.

# 3 Radio transmission and reception

This chapter describes the techniques that are used for radio transmission and reception between the mobile and the radio access network. The first section reviews the use of wideband code division multiple access for transmission and reception in release 99. It concentrates on the air interface's physical layer, which is where most of the important processes take place, but it also notes the procedures used in higher layers. We then describe a technique known as high speed packet access, which has been progressively introduced from release 5 with the aim of increasing the rate at which data can be transferred. Finally, we discuss the performance of UMTS, by noting the peak and average data rates that can be achieved, and the advantages and disadvantages that CDMA has compared with other multiple access techniques.

The material in this chapter is more technical than that in later chapters of the book. However, it is unnecessary to take it all in on a first reading, as most of the chapter is self-contained. Instead, a basic understanding of Section 3.1 will be enough for Chapters 4, 5 and 6.

## 3.1 Radio transmission and reception in release 99

In this first section, we will describe the techniques used for radio transmission and reception in release 99. This is probably the most important part of the system: it is crucial for the delivery of high data rates to the user, and it is the part that has changed the most since the days of GSM.

The action takes place in the air interface's transport protocols, which are illustrated in Figure 3.1. In the transmitter, the radio link control and medium access control protocols handle tasks such as retransmissions and control of the transmitted data rate. The physical layer then manipulates the data in three stages. In the first stage, the data are processed one bit at a time, to carry out tasks such as error correction coding. In the second stage,

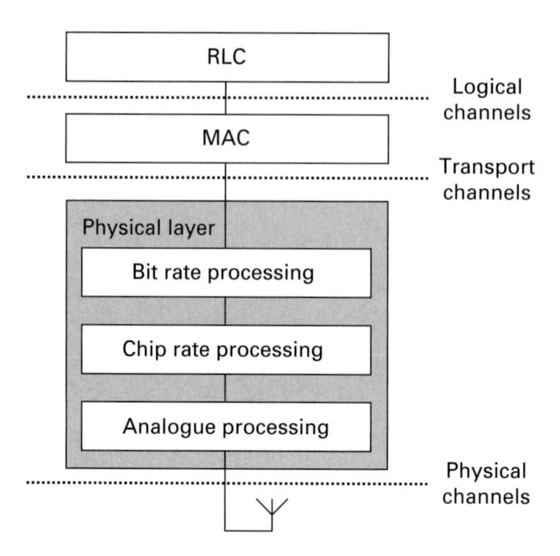


**Figure 3.1** Architecture of the air interface's transport protocols, showing the internal architecture of the physical layer.

the coded bits are divided into shorter units called *chips*, and the chips are processed one at a time using the techniques of CDMA. Finally, the chips are converted from digital to analogue form for transmission over the air interface.

As an example, we can illustrate the division of bits into chips by describing how the mobile might transmit a voice call. At the point where the data enter the physical layer, the data rate is typically 12.2 kbps. Using error correction coding and another process called rate matching, the bit rate processor increases the bit rate by a factor between 2 and 3, here to 30 kbps. The chip rate processor then divides each coded bit into 128 chips, to produce a chip rate of 3.84 million chips per second (Mcps). The same chip rate is used throughout UMTS FDD mode, but the other numbers can vary from one data stream to another, and between the uplink and downlink.

We will start by discussing the chip rate processor, which is where most of the new features of UMTS are located. After a detour to the analogue processor, we will move upwards through the protocol stack and finish at the RLC.

### 3.1.1 Principles of CDMA in UMTS

In this section, we will describe the basic principles of CDMA. The treatment starts with the downlink, using a greatly simplified network that contains just one mobile and one cell. We then develop things to include multiple mobiles and multiple cells, and finally look at the differences between the downlink and the uplink. The description is somewhat simplified: we will note a few differences in the actual implementation in the sections that follow.

Figure 3.2 shows the chip rate processing that is carried out on the downlink, in a network containing one mobile and one cell. At the top of the figure, the base station's chip rate transmitter is handling a stream of bits that it wishes to send to the mobile. The base station assigns the mobile a code that is known either as a *channelisation code* or a *spreading code*: this is made of chips and has a length equal to the

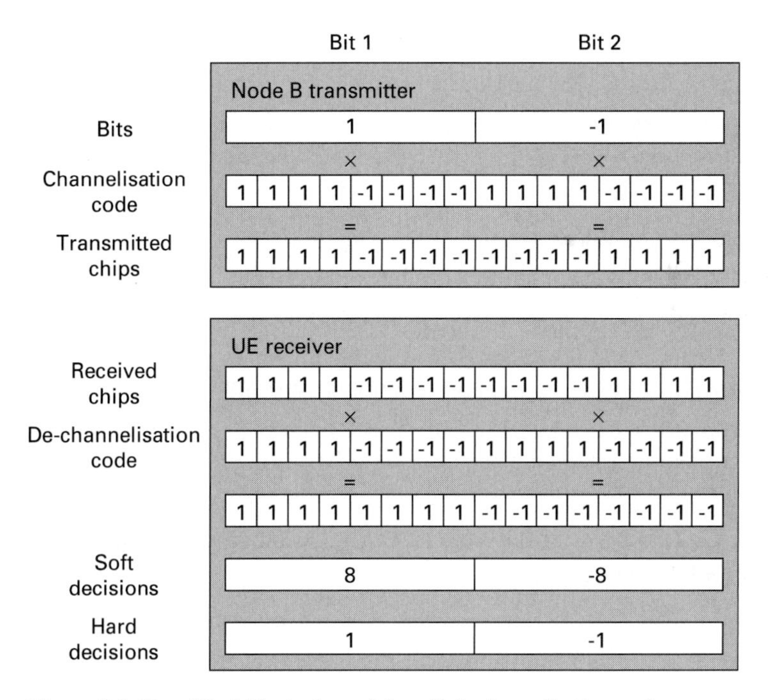


**Figure 3.2** Simplified illustration of downlink channelisation and de-channelisation, from a base station to a single mobile.

bit duration, so that the code is repeated every bit. It then multiplies the symbol representations of the bits and chips together, and sends the resulting chips to the analogue processor for transmission.

The bits and chips both have values of 0 and 1, but we have represented them using binary phase shift keyed (BPSK) symbols of +1 and –1. We will continue to use this symbol representation in the discussion that follows, but we will normally describe the quantities as bits and chips rather than symbols, to make it easier to distinguish the bits and chips from each other.

In UMTS FDD mode, the chip rate is fixed at 3.84 Mcps, so the chip duration is about 0.26 μs. The number of chips per bit is called the *spreading factor*: in this example, the spreading factor is 8. The bit rate equals the chip rate divided by the spreading factor, so here the bit rate is 480 kbps. (Note that error correction has already been applied to these bits, so the underlying information rate is typically one third to one half as great.)

If we ignore problems like noise and propagation loss, then the mobile's chip rate processor receives an exact replica of the transmitted chips. We now assume that the base station has previously told the mobile about the channelisation code that it will use, so that the mobile can use this information to undo the effect of channelisation. It does this by multiplying the incoming chips by the channelisation code (or, more accurately, by multiplying their symbol representations together).

The mobile now has to convert the chips into bits, which it does by adding together the chips that comprise each bit. The result is a set of *soft decisions*, each of which has a sign corresponding to the mobile's best estimate of the transmitted bit, and a magnitude corresponding to the mobile's confidence in that estimate. Finally, the mobile converts the soft decisions into *hard decisions* by taking the sign. We can see that the mobile has recovered the original bits: it has done this because the process of channelisation was reversed in the mobile's de-channelisation stage.

Figure 3.3 shows what happens if there are two mobiles in the cell. Here, the base station assigns a different channelisation code to the second mobile, with the condition that the two codes must be *orthogonal*: if we

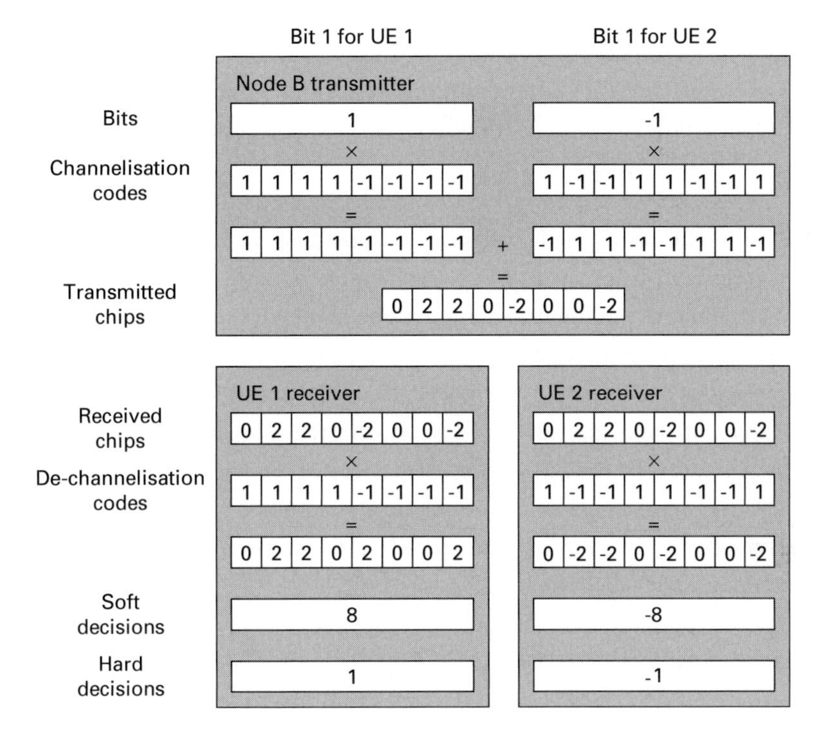


**Figure 3.3** Simplified illustration of downlink channelisation and de-channelisation, from a base station to two mobiles.

multiply them together chip-by-chip and add up the results, the total must be zero. The base station multiplies the incoming bits by their respective channelisation codes as before, and then adds the two streams together, chip-by-chip. The transmitted chip stream contains signal levels of +2, 0 and –2, where each chip has contributions for both of the two mobiles. The receive processing is unchanged: each mobile multiplies the incoming stream of chips by its own channelisation code, adds together the chips that comprise each bit, and calculates a set of hard decisions.

In the figure, the two mobiles have successfully recovered the bits that were intended for them, despite the fact that the transmitted stream contained information for both mobiles. This works because the two channelisation codes are orthogonal. Let's consider, for example, what happens to the symbol that was intended for mobile 1, when it is received by mobile 2. The symbol was transmitted using mobile 1's channelisation code, so when it is processed using mobile 2's de-channelisation code, the result is zero. It therefore has no effect on the soft decisions that are made by mobile 2.

The number of orthogonal codes available is equal to the spreading factor: eight orthogonal codes at a spreading factor of eight, for example. Figure 3.4 shows the codes that are actually used by UMTS. We can think of the figure as a family tree in which each channelisation code has two children, one made by repeating it, and the other by repetition and inversion. The codes on each spreading factor are all mutually orthogonal, while codes on different spreading factors are orthogonal too, so long as they are not ancestors or descendants of each other. The spreading factors are implicitly restricted to integer powers of 2: in release 99, we only use spreading factors from 4 to 512 on the downlink, and 4 to 256 on the uplink.

What happens if there is more than one cell? The channelisation code tree can only accommodate a limited number of mobiles, so we want to re-use it in every cell. This causes a problem if two nearby cells are transmitting on the same frequency and the same channelisation code, because of cross-talk between the two transmissions. We solve the problem by introducing a second set of codes, known as *scrambling codes*, and

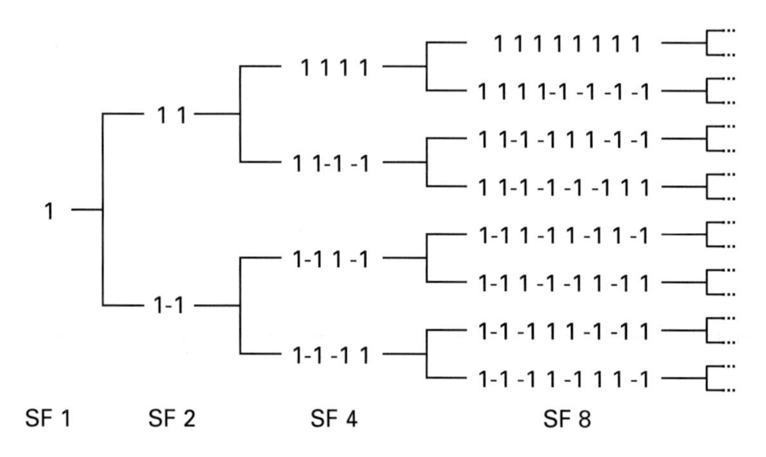


**Figure 3.4** Channelisation codes used by UMTS. (Adapted from 3GPP TS 25.213.)

labelling each nearby cell with a different scrambling code. The scrambling codes are made of chips, but they have a much longer repetition period than the channelisation codes: 10 ms, which is 38 400 chips. In UMTS, there are enough scrambling codes to label 512 different cells. This number is large enough that cells on the same scrambling code are a large distance apart, and the cross-talk between them is minimal.

Ideally, the scrambling codes would be orthogonal to each other, but we have unfortunately used up all the orthogonal codes in making the tree of channelisation codes. Instead, we do the next best thing: we create them using a pseudo-random number generator, which makes the scrambling codes *uncorrelated*. If we multiply two scrambling codes together and add up the results as before, then the total is zero on the average, but it is not identically zero.

The effect of using uncorrelated codes is shown in Figure 3.5, which has four cells with different scrambling codes, each transmitting to a

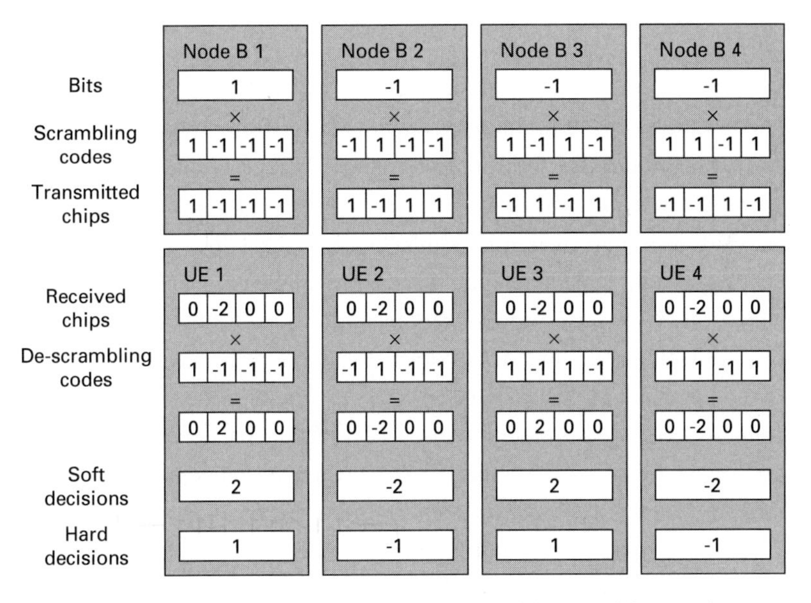


**Figure 3.5** Simplified illustration of scrambling and de-scrambling on the downlink. Each mobile receives a signal from its corresponding Node B, and interference from the other three. The processes of channelisation and de-channelisation are omitted.

different mobile. In this figure, the spreading factor is four, so we only show the first four chips from each scrambling code. The channelisation codes are left out for clarity: we could choose channelisation codes of 1111 throughout, which would leave all the other numbers unchanged. We also assume that each mobile receives equally strong signals from the four cells; this is a rather artificial situation, but it serves to illustrate the point.

Each cell applies its scrambling codes by a chip-by-chip multiplication, and the process is reversed in the mobile receiver. Because the scrambling codes are uncorrelated and not orthogonal, the mobiles receive some interference from neighbouring cells. This perturbs the soft decisions away from their expected values, and occasionally (as in the case of mobile 3) causes errors in the hard decisions. The receiver can correct most of these errors later on using error correction and retransmissions, but some of them will leak through and degrade the performance of the application. As we will discuss in Section 3.3, the interference and the resultant errors are a very important issue in UMTS, and will ultimately limit the capacity of the system.

In the uplink, the processing steps are exactly the same, but the channelisation and scrambling codes are used differently. The reason is that, in FDD mode, the transmissions from different mobiles are not time synchronised in any way. This simplifies the design of the system, but it has a disadvantage: it is impossible to distinguish the mobiles by the use of different channelisation codes, because those codes are only orthogonal if they are time synchronised with each other. Instead, the network distinguishes different mobiles by assigning different scrambling codes to them, whichever cell they are in. (The total number of uplink scrambling codes is about four million.) The channelisation codes are only used for two purposes: to set the data rate by means of the spreading factor, and to distinguish different transmissions from a single mobile.

We will discuss the exact implementation of the chip rate transmitter and receiver in Sections 3.1.3 and 3.1.5 below, with the underlying analogue processing coming between them in Section 3.1.4. First, however, we need to digress to an issue that is important for chip rate processing in CDMA: power control.

### 3.1.2 Power control

*Power control* is best introduced by describing a problem on the uplink known as the *near far effect*. Let's assume that a base station is receiving signals from two mobiles, one close to the base station and one far away. The base station's analogue-to-digital (A/D) converter has a limited dynamic range (just a few bits), so if the two signals are transmitted with the same power, then the nearby one can saturate the A/D converter while the distant one is hardly resolved at all. The base station will then be unable to hear the distant mobile. This is not a problem for systems based on FDMA or TDMA, where the signals can be separated before digitisation using their carrier frequencies or arrival times, but it is an important issue for CDMA.

The solution depends on the physical channel of interest. For the dedicated physical data channel (DPDCH) and the dedicated physical control channel (DPCCH), the answer lies in fast closed loop power control (Figure 3.6). Every two-thirds of a millisecond, the base station measures the *signal-to-interference ratio* (SIR) that it receives from each

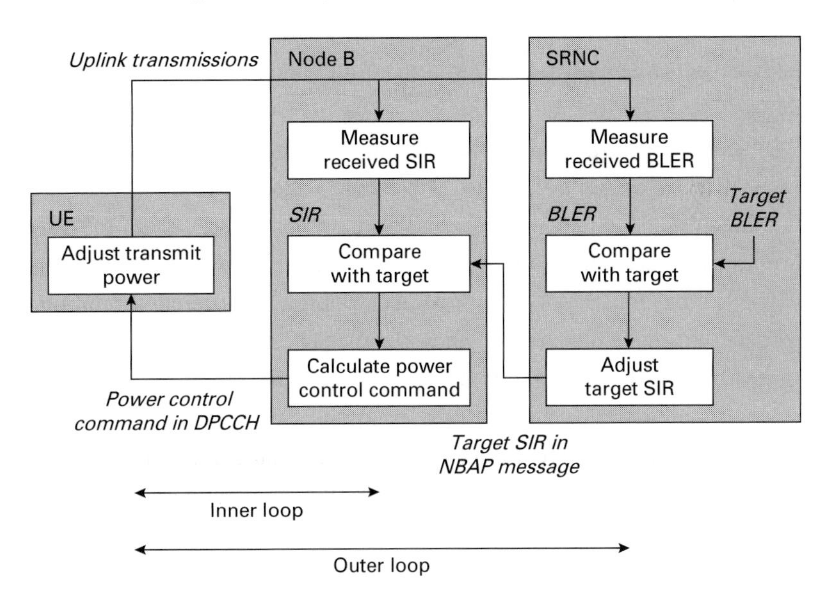


**Figure 3.6** Operation of fast closed loop power control for the DPDCH and DPCCH uplink.

mobile, and compares it with a target SIR. It then uses the result to compute a power control command, which it sends to the mobile on the downlink DPCCH.

The power control command is calculated as follows. If the received SIR is below the target, then the base station tells the mobile to increase its transmit power, typically by 1 dB. Similarly, if the SIR is above the target, then the base station tells the mobile to decrease its transmit power. As a result, the mobile adjusts its transmit power up or down every two-thirds of a millisecond, so as to hold the base station's received SIR close to the target value.

The target signal-to-interference ratio is itself adjusted using an outer loop, as follows. Over a longer timescale, the serving RNC measures the error rate in the received signal using a quantity called the *block error ratio* (BLER). It then compares this with a target BLER that is associated with the data stream. (We will define the BLER in Section 3.1.6, and list some suitable target values.) If the BLER is greater than its target value, then the serving RNC sends a signalling message to the Node B, in which it tells the Node B to adjust the target SIR upwards. The Node B uses the target SIR in its inner loop, so it immediately tells the mobile to increase its transmit power. This causes the error rate at the serving RNC to fall.

If the BLER is less than the target value, then the serving RNC tells the Node B to adjust its target SIR downwards, and the Node B tells the mobile to reduce its transmit power. This clearly increases the number of errors in the received data stream, but it brings two important benefits. First, by reducing the mobile's transmit power, we increase its battery life. Second, we reduce the amount of interference at the base station receiver, so we increase the capacity of the cell. We can sum up the effect of the control loop by saying that the mobile's transmissions are at the lowest power consistent with satisfactory reception of the signal.

Fast power control brings one more advantage, which is shown in Figure 3.7. If a mobile moves into a fade, then the base station tells it to increase its transmit power, so as to keep the received signal-to-interference ratio at roughly the same level. This reduces the amount of fading in the received signal, so it reduces the error rate. As shown in

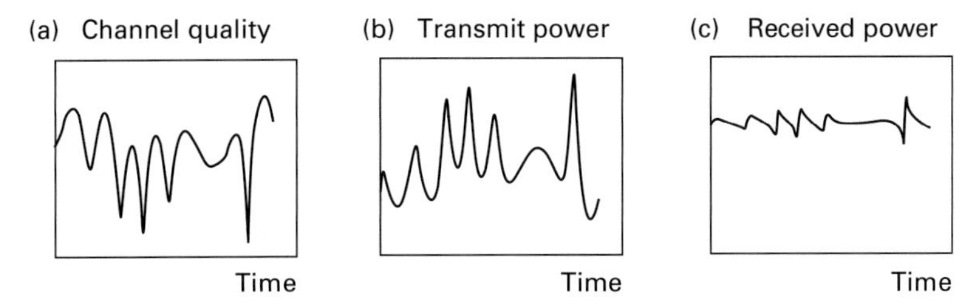


**Figure 3.7** Illustration of how fading can be reduced by the use of fast power control. (a) Received signal-to-interference ratio in the absence of fast power control. (b) Transmitted power when using fast power control. (c) Received signal-to-interference ratio in the presence of fast power control.

the figure, we cannot remove fading altogether because of issues such as time delays in the control loop, but the improvement is an important one nevertheless.

For these reasons, fast power control is used for the downlink DPDCH as well as for the uplink. On the downlink, fast power control is applied independently to every data stream, so the base station sends high power signals to mobiles that are far away, and low power signals to mobiles that are nearby.

Power control is also important for the physical random access channel (PRACH). The mobile typically uses this channel for short transmissions such as small bursts of packet data or discrete signalling messages. It is harder to control the power of the PRACH, because the mobile has to work out its transmit power without the help of a continuously running control loop. The solution is shown in Figure 3.8. The mobile starts by sending a preamble made of short bursts. The first burst is at a low enough power that the base station is unlikely to hear it, while each subsequent burst is typically 2 or 3 dB stronger than the previous one. Once the base station hears the mobile, it sends a reply on the acquisition indicator channel (AICH). The mobile can then send its uplink data, confident that its transmit power is strong enough for the base station to hear it, but not so strong as to cause unnecessary interference to the base station. The messages are very short (either 10 or 20 ms long), so we can assume that the required transmit power will not change significantly before the end of the message. The mobile therefore transmits the message with a constant power.

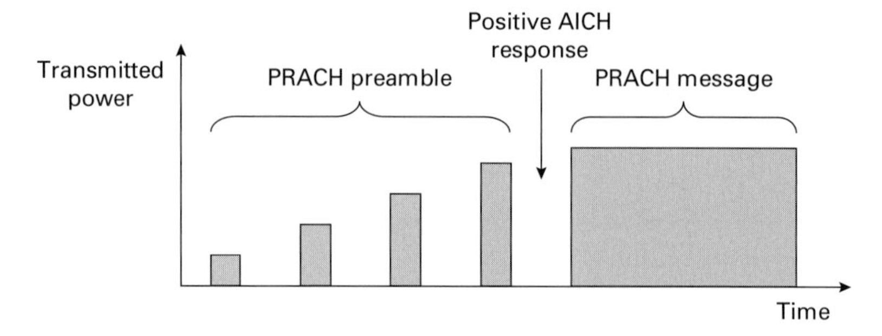


**Figure 3.8** Operation of open loop power control for the PRACH.

The remaining physical channels are downlink-only channels that are designed to cover the whole of the cell. They are usually sent at a constant power, without any further adjustment.

### 3.1.3 Chip rate transmitter

In this section, we will look at how channelisation and scrambling are actually implemented in UMTS. We will just do this for the DPDCH and DPCCH: the other physical channels are broadly similar, but have a number of differences in the detail.

Figure 3.9 shows how the two channels are organised, at the point where they enter the chip rate transmitter. In both the uplink and downlink, the information is laid out in 10 ms *frames*, which are divided into 15 *slots* of ⅔ ms each. Within each slot, the DPDCH and DPCCH are sent in parallel on the uplink but in series on the downlink. The DPDCH contains data, while the DPCCH has four different control fields. The *pilot* information is a known stream of bits, which help the rake receiver (Section 3.1.5) to process the incoming data stream. The *transport format combination indicator* (TFCI) signals the type of information that the DPDCH frame contains, such as voice, signalling or both. The *transmit power control* (TPC) bits are the power control commands described above. The *feedback information* (FBI) bits are less important than the others: they are only used on the uplink, and control an optional process called closed loop downlink transmit diversity.

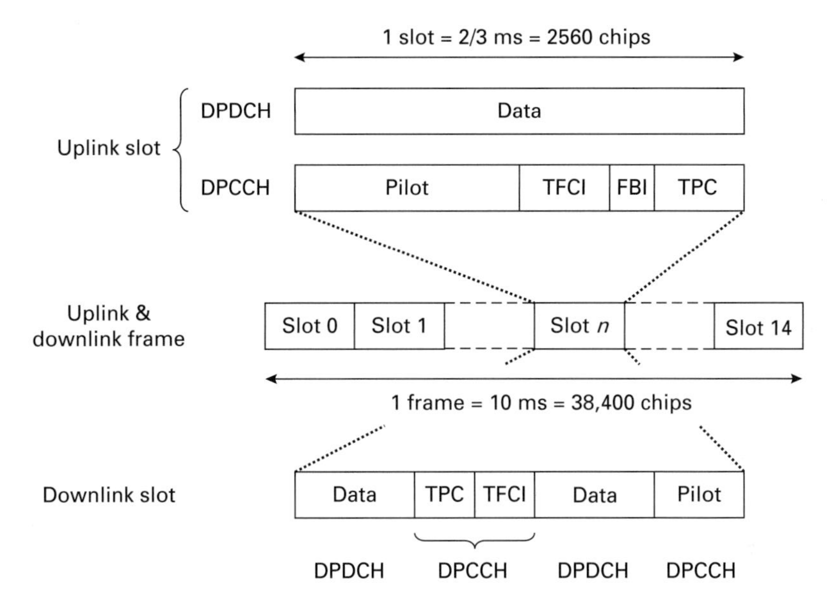


**Figure 3.9** Frame and slot structure for the DPDCH and DPCCH. (Adapted from 3GPP TS 23.211.)

Figure 3.10 is a block diagram of the uplink transmitter. As shown in the figure, the uplink normally uses one DPDCH and one DPCCH. If the mobile is sending more than one information stream (such as voice and signalling, for example), then the streams are normally multiplexed onto the same DPDCH, with the TFCI indicating the type of information being sent. The mobile converts the bits into BPSK symbols, and multiplies them by a channelisation code. The DPDCH uses a spreading factor that depends on the bit rate, while the DPCCH uses a fixed spreading factor of 256. The specifications define the exact choice of channelisation codes.

The mobile then determines how much power is transmitted in the two channels, by multiplying the data by suitable scaling factors. To transmit a high power signal, for example, it might convert symbols of ±1 into ±10, while for a low power signal it might convert them into ±0.1. These scaling factors are actually applied in two steps. The first step comes after channelisation, where the mobile applies digital scaling factors to set the relative transmit powers of the two channels. The second step comes in the analogue transmitter, which can handle a much bigger dynamic range.

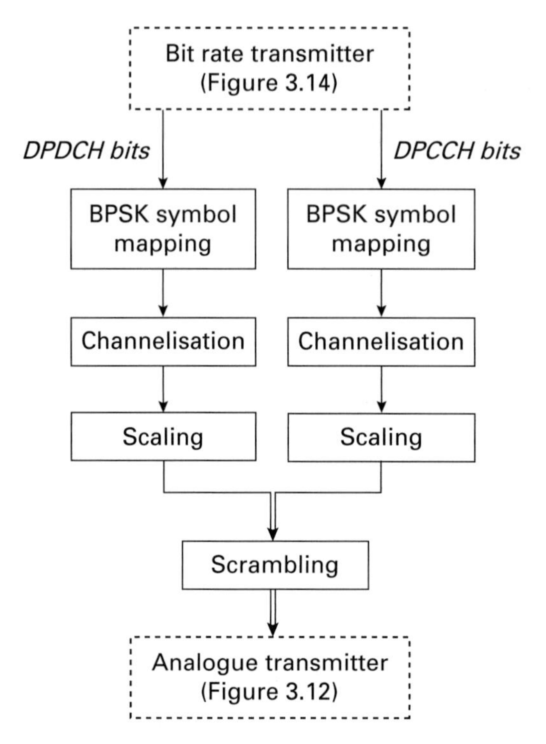


**Figure 3.10** Block diagram of the uplink chip rate transmitter for the DPDCH and DPCCH, in the usual case of transmission on a single DPDCH.

There, the mobile adjusts the absolute transmit power using the power control commands received from the Node B.

Between these two steps is the scrambling operation, which is slightly more sophisticated than the process described earlier but works on the same basic principles. The DPDCH and DPCCH are treated as the real and imaginary parts of a complex number, otherwise known as the in-phase and quadrature components (I and Q). The complex number is indicated in the figure by a double line. The scrambling code is a complex number as well, and the scrambling operation includes a complex multiply.

This process works fine at information rates up to about 384 kbps, but at higher rates more data channels are required. In release 99, a mobile can use up to six uplink data channels with a spreading factor of four, to give a maximum data rate of about 2048 kbps. Three of the channels are transmitted on the in-phase component and three on the quadrature component, which ensures that the data streams are still orthogonal to each other, even though each channelisation code is used twice. In practice, it is very unusual for mobiles to transmit at such high data rates in this way. Instead, they normally use the techniques of high speed uplink packet access that are described in Section 3.2.

The downlink (Figure 3.11) is handled a bit differently. At the start of the process, the bits for each mobile are taken two at a time and converted into quadrature phase shift keyed (QPSK) symbols, which are treated as complex numbers with values of $\pm 1 \pm i$. After channelisation, the symbols are scaled using the power control commands received from their respective mobiles. The symbols for different mobiles are then added together, scrambled, and sent to the analogue transmitter as before. The maximum data rate on the downlink is again about 2048 kbps per mobile.

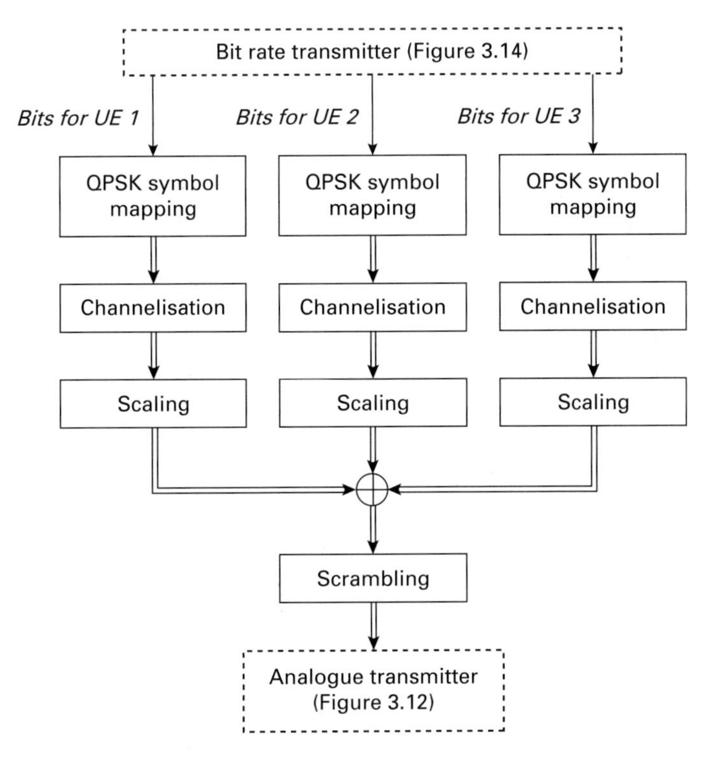


**Figure 3.11** Block diagram of the downlink chip rate transmitter for the DPDCH and DPCCH.

The reason for the difference between the uplink and the downlink is as follows. On the downlink, the channelisation codes are a scarce resource. The DPDCH and DPCCH are therefore transmitted in series, on the same channelisation code, which uses the codes in the most efficient way. This technique has one disadvantage, however: if the two channels require a different transmit power, then serial transmission makes the transmitted power vary at a frequency of $(2/3\,\text{ms})^{-1}$, which is 1.5 kHz. This is an audio frequency, so it causes interference to radios and other audio devices. On the uplink, we can avoid any risk of interference by transmitting the DPDCH and DPCCH in parallel, to keep the transmit power constant. On the downlink, interference is usually much less of a problem for two reasons: the transmitter is further from any susceptible devices, and the transmissions to multiple mobiles keep the total power much the same.

### 3.1.4 Analogue processing

There are various ways of implementing the analogue processor. In this section, we will just give an outline of one such implementation, to highlight the principles that are involved.

Figure 3.12 shows an analogue processor that uses direct conversion in both the transmitter and the receiver, with no intermediate frequency. The I and Q streams are first passed through a particular type of digital filter known as a *Nyquist filter*, which smoothes the transitions between successive chips, without disturbing the waveform at the sampling times the receiver will eventually use. The effect of the filter is to reduce the signal bandwidth in the frequency domain, which ensures that the signal can be transmitted and received in the available bandwidth of 5 MHz. The Nyquist filter used in UMTS is a *raised cosine* filter, but it is actually implemented as two separate *root raised cosine* filters. One of these is in the transmitter to reduce the transmitted signal bandwidth, and the other is in the receiver to reduce interference from signals on nearby carriers.

In the transmitter, the filtered I and Q streams are converted from digital to analogue, and are amplified and filtered at baseband. They are then converted to radio frequency (RF) by mixing them with cosine and sine wave carriers, and added together. The signal is then amplified and

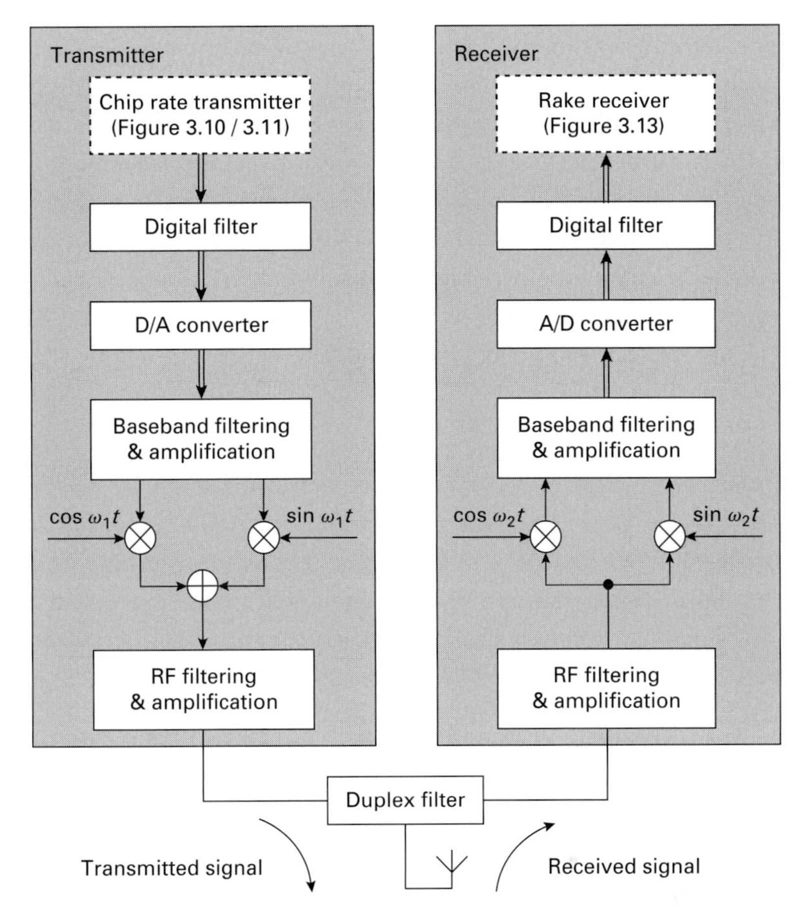


**Figure 3.12** Implementation of the analogue processor using direct conversion in both transmitter and receiver.

filtered at RF before being sent to the antenna, by way of a duplex filter that isolates the transmit and receive chains from each other.

In the receiver, the processing is reversed. The incoming signal is amplified and filtered at RF, before being mixed with cosine and sine wave carriers that convert it to baseband. After low pass filtering and a further stage of amplification, the baseband signals are digitised and passed to the root raised cosine filter.

Before we end this section, it is worth noting a problem that affects the receiver's processing: the phase angle of the received signal is at this stage completely unknown. If, for example, we move the receive antenna through half a wavelength of the carrier (typically 75 mm in UMTS), then the received waveform is inverted, and we change the signs of all the chips received on I and Q. Even worse, if we move the antenna through a quarter of a wavelength, then (ignoring a sign change) we interchange I and Q. Generalising, we can say that the I and Q signals in the receiver are not the same as the ones in the transmitter, but instead are linear combinations of them:

$$\begin{aligned} I_{\text{received}} &= I_{\text{transmitted}} \cos\phi + Q_{\text{transmitted}} \sin\phi \\ Q_{\text{received}} &= Q_{\text{transmitted}} \cos\phi - I_{\text{transmitted}} \sin\phi \end{aligned} \tag{3.1}$$

for some phase shift $\phi$. The situation is worse in a multipath environment, where there are several copies of the received signal, each of which has a different value of $\phi$. If we are to process the received signal successfully, then we have to measure and remove those phase shifts, so as to map the received chip streams onto the ones that were originally sent. This will happen as part of the rake receiver processing, which we will now discuss.

### 3.1.5 Rake receiver

The receiver's chip rate processor uses a device known as the *rake receiver*, which is shown in Figure 3.13. We will describe it by referring to the downlink, but the treatment refers equally well to the uplink.

The aim of the receiver is to reduce the amount of fading in a multipath environment, by applying diversity processing to the incoming rays. We therefore assume that the mobile is receiving two distinct rays from the base station. In the receiver, we set up two processing streams that are known as *fingers*: the two fingers receive exactly the same information, but fingers 1 and 2 are configured to process the information arriving on rays 1 and 2 respectively.

At the start of finger 1 is a delay buffer, which is configured so that the mobile's de-scrambling code is time aligned with the chips that are arriving on ray 1. This implies that, at the output of the de-scrambling stage, finger 1 contains a de-scrambled signal from ray 1. However,

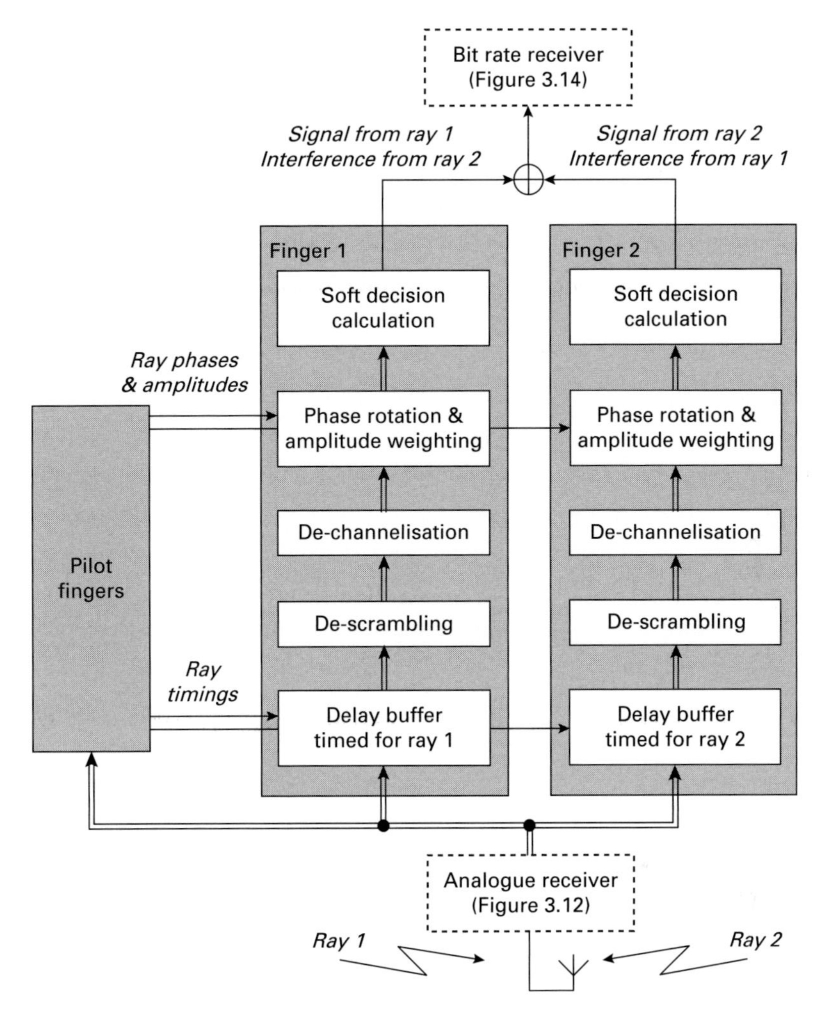


**Figure 3.13** Block diagram of the rake receiver, for the case where the receiver is processing two distinct rays.

finger 1 is also receiving chips that arrive on ray 2. These have a different arrival time, so they are misaligned with the mobile's de-scrambling code, and they generate a small amount of interference. The processing in finger 2 is the same, except that its delay buffer is configured to align the de-scrambling code with the chips arriving on ray 2. The effect is that finger 2 produces a signal from ray 2 and interference from ray 1.

After de-scrambling, the data are de-channelised and converted to soft decisions as before. The calculation of hard decisions is left until much later, in the bit rate receiver.

We also need to remove the phase shifts $\phi$ that we noted above. To do this, the receiver processes the pilot bits on either the DPCCH or the downlink's common pilot channel (CPICH), using separate rake fingers that are configured using the channelisation code for the corresponding channel. The transmitted pilot bits are defined in the 3GPP specifications, and are received with the same phase shift as all the other data. By comparing the received pilot bits with the expected ones, we can measure the value of $\phi$ for each ray and send the measurement to the corresponding rake finger, which removes the phase shift from the data stream. This process allows us to add together the two sets of soft decisions at the end of the rake receiver, without any risk of destructive interference between them. We also apply a weighting factor to each data stream, which is computed using the amplitude of the corresponding pilot signal. The weighting factor ensures that the signal-to-interference ratio, after the two data streams are added together, is as large as possible.

The receiver has a couple of other tasks, which are not shown in the figure. First, a component known as the *searcher* runs the cell search procedure that we will describe in Chapter 4, so as to discover when new rays appear and old rays disappear. The searcher sends information about the rays it finds to the rest of the receiver, which responds by setting up new rake and pilot fingers, and tearing down the ones that are no longer required. Second, there are extra components associated with the pilot fingers that measure the arrival time of each ray, by locking on to the expected sequence of pilot bits and adjusting for any timing changes.

The effect of the rake receiver is that we have processed the two rays independently, in much the same way as for receive antenna diversity, so as to reduce the amount of fading in the received signal. However, the process only works if the difference in the arrival times of the two rays is greater than the chip duration. (If they arrive closer together, then the rake receiver cannot distinguish them, so they are handled by a single finger and cause fading as before.) This implies that the path difference between the rays must be greater than about 80 m, which is the distance travelled by a radio signal over the duration of one chip. This is easy to achieve in an urban macrocell that contains lots of widely spaced reflectors, but it is harder in a rural environment or a microcell, and is usually impossible in an indoor picocell.

If there are several distinct rays, then the receiver sets up one finger for each ray, starting with the highest power ray and moving on to weaker ones. Two issues limit the number of fingers used: the processing power in the receiver, and the diminishing returns that result from processing weaker and weaker rays. If there are several fingers, then they start to look like the prongs on a garden rake, which explains the receiver's name.

Finally, note that the receiver does not care where the two rays in Figure 3.13 come from. On the downlink, for example, the rays can come from different cells, so long as the mobile uses different de-channelisation and de-scrambling codes in the two fingers. This means that a mobile in soft handover can process the rays from different cells by adding them together, in the same way that it processes rays from a single cell. As a result, most of the processing needed for soft handover is just a by-product of the processing in the rake receiver.

### 3.1.6 Bit rate processing

We will now move upwards through the physical layer and look at the bit rate processor. Figure 3.14 summarises the processing that it carries out: there are more blocks in practice and the details are different on the uplink and downlink, but the figure covers all the important issues.

In the transmitter, data arrive on the transport channels, in the form of physical layer service data units that are known as *transport blocks*. These have a duration known as the *transmission time interval* (TTI), which can take a value from 10 to 80 ms (1 to 8 frames) depending on the particular data stream. Roughly speaking, each transport channel carries a single stream of information such as voice, signalling or packet data.

Skipping over the first process for the moment, the second transmit process is error correction coding, which we introduced in Chapter 1. In this process, we represent the information bits using codewords, so as to double or treble the number of transmitted bits. The algorithm used is

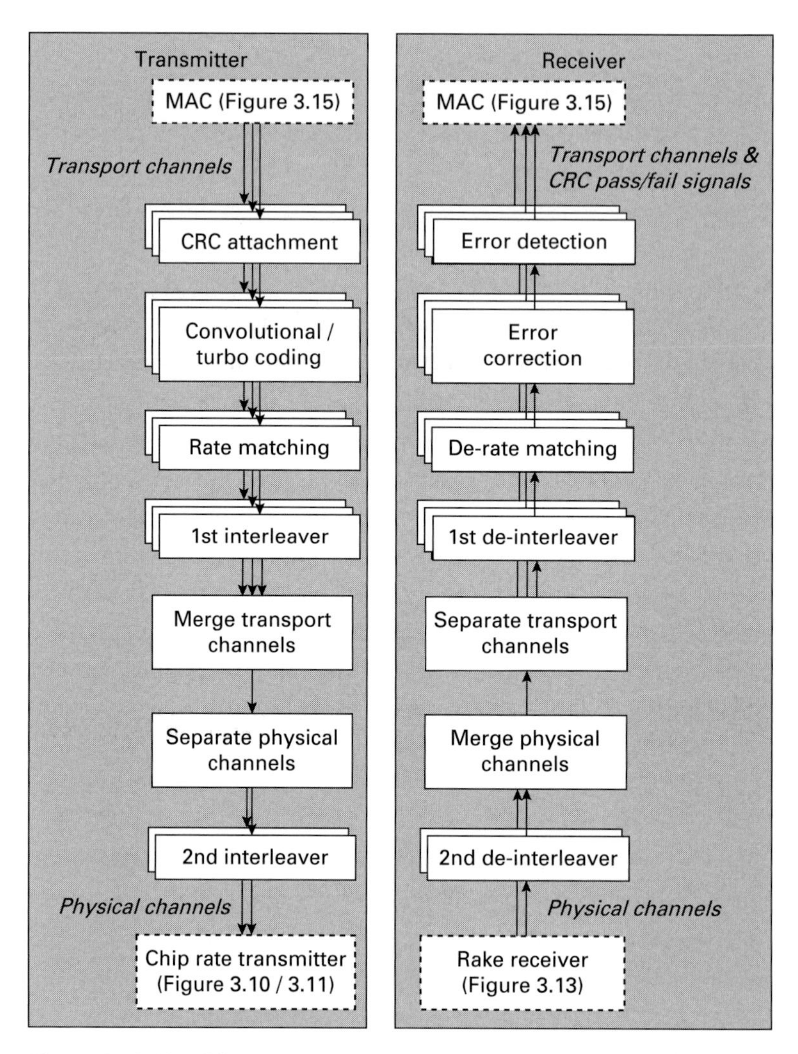


**Figure 3.14** Simplified block diagram of the bit rate processor. (Adapted from 3GPP TS 25.212.)

either *convolutional coding* or *turbo coding*: the first is used at low bit rates, while the second gives better results at bit rates above about 32 kbps. The receiver uses the extra information for error correction, which greatly reduces the error rate. The receiver's hard decisions are usually computed as part of the error correction stage, so the soft decisions run as far as the input to it.

Unfortunately there is no such thing as a perfect error correction algorithm, so a few errors will leak through. These are handled by the blocks we skipped over: *cyclic redundancy check* (CRC) *attachment* in the transmitter, and *error detection* in the receiver. In the transmitter, we use the bits in the transport block to compute and append a few extra bits that are known as the CRC. In the receiver, we examine the transport block and CRC to see if they are still consistent. If they are, then everything is fine; if not, an error has occurred. The action we then take depends on the nature of the data stream. For real time streams such as voice, timely delivery is more important than accuracy, so we just tell the application software about the error and let it decide what to do. (It might estimate the erroneous block's contents using the previous block, for example.) For non-real time streams such as emails or web pages, we send a signal to the RLC protocol, which asks for a retransmission.

The percentage of transport blocks that fail the CRC is the block error ratio (BLER) that we introduced earlier. Voice and video streams run happily with BLERs up to around 1% and 3% respectively, and these are typical target BLERs for those streams. For packet data, the target BLER is usually around 10%: we can tolerate such a high figure because the errors are corrected later on using retransmissions.

Like the error correction stage, error detection is not perfect: a transport block can occasionally pass the CRC, even though it contains bit errors. These residual bit errors leak through to the application, and in some cases (emails and web browsing, for example) have the potential to be a serious problem. By using enough CRC bits, the network can reduce the residual bit error ratio to a level the application can tolerate.

*Rate matching* deals with a problem that is implicit in our description so far: the transmitter has to handle any bit rate that the application throws at it, but the physical layer only uses spreading factors that are an integer power of 2. If the input stream contains too many bits for the target spreading factor, then the transmitter removes a few of the bits by a process called *puncturing*. For example, the error correction algorithm might represent a set of 4 coded bits using a 12-bit codeword. The puncturing algorithm might match the chosen spreading factor by removing one of these, reducing the length of the codeword to 11 bits. The receiver runs the same algorithm, so it can work out the places where the punctured bits occurred: it then inserts soft decisions of 0 at the appropriate places, and leaves the error correction stage to recover the information bits. If the transmitter has too few bits, then it either duplicates some of them by *repetition*, or leaves gaps in the transmitted data stream.

*Interleaving* solves another problem. The coding stage distributed the information from a single input bit over several coded bits, but the coded bits can all be ruined if they arrive within a single fade. Interleaving redistributes the coded bits across the transport block, to ensure that at least some of the coded bits from each information bit arrive correctly. As shown in the figure, interleaving is actually carried out in two stages: the second interleaver distributes the coded bits over a 10 ms frame, and the first interleaver does some extra work if the TTI is longer than a frame.

Between the two interleavers, the transport blocks are broken apart into frames, and the frames from the different transport channels are multiplexed together. If necessary, the information can then be distributed amongst multiple physical channels, although this is usually only done if the data rate is too high for one channel to cope.

### 3.1.7 Medium access control protocol

Figure 3.15 shows the architecture of the medium access control (MAC) protocol. There are two main components: the MAC-d handles dedicated channels, and the MAC-c/sh handles common channels. The network has one MAC-c/sh per cell, which is implemented in the cell's controlling RNC, and one MAC-d per mobile, implemented in the mobile's serving RNC.

The link between the MAC-d and MAC-c/sh supports the transmission of dedicated logical channels using common transport and physical channels. These are handled as follows. On the downlink, the base station labels the transport blocks using the identity of the target mobile. It then transmits them on the forward access channel (FACH) and the secondary common control physical channel (SCCPCH). When a mobile reads the channel, it processes the transport blocks if the label matches its identity, and discards them otherwise. A similar process takes place on the uplink.

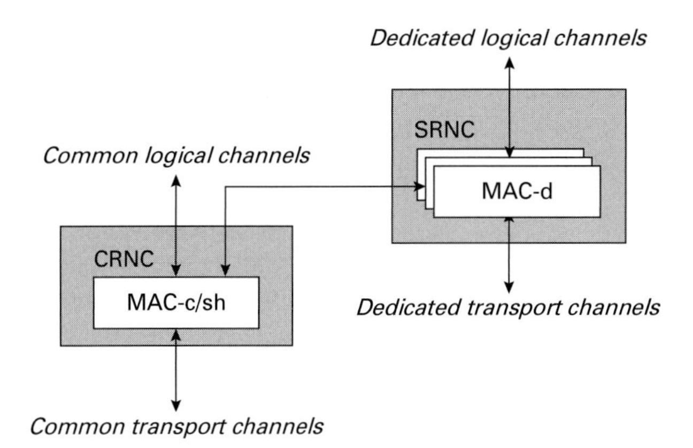


**Figure 3.15** Architecture of the medium access control protocol. (Adapted from 3GPP TS 25.321.)

The MAC has many functions, but its main role is to decide how many bits are sent per transmission time interval from each of the transport channels. The process works as follows. When the data streams are first set up, the network associates each transport channel with a table known as a *transport format set* (TFS). This lists the allowed transport block sizes for the channel in question, and for each block size, the number of transport blocks that are transmitted in parallel. Each row in the table is labelled using a *transport format indicator* (TFI).

The network also sets up a *transport format combination set* (TFCS). This defines how the transport formats from different channels can be combined, leaving out any combinations which are mutually inconsistent or which would lead to too high a data rate. Each combination is labelled using a *transport format combination indicator* (TFCI). Finally, the network sends this information to the mobile, using separate tables for the channels on the uplink and downlink.

The network also associates each transport channel with a transmission priority. This is set to a high value for a real time stream such as voice, and low for streams that can tolerate delays such as signalling or non-real time packet data.

Every TTI, the MAC finds out how much data are waiting to be transmitted, chooses a row from the transport format combination set, and sends the data to the physical layer. The algorithm in the mobile is well defined: grab as much data as you can from the highest priority channel in a way that is consistent with the transport format combination set, and then move on to the next one. The network has more room for manoeuvre and has to juggle transmissions to all the mobiles in the cell, but would use a broadly similar technique. The MAC also sends the corresponding TFCI to the physical layer, which sends it to the receiving device, for example using the DPCCH (Figure 3.9). This tells the receiver what is going on, so that it can process the incoming data correctly.

Table 3.1 shows an example. Here, the mobile is transmitting two transport channels, both of which have three transport formats (Table 3.1a). Channel 1 adjusts its bit rate by adjusting the transport block size, while channel 2 uses the number of transport blocks. The TFCS could have up to nine transport format combinations, but in this example (Table 3.1b), only six are allowed because the others lead to too high a bit rate. Every TTI, the mobile inspects its transmit buffers, chooses a row from the table and sends the data to the physical layer, along with a note of the selected TFCI.

### 3.1.8 Radio link control protocol

The last protocol that we will cover is the radio link control protocol. The RLC has three modes of operation, one of which is chosen for each individual data stream. The simplest is *transparent mode* (TM), which passes the transmitted data directly from input to output with little or no further processing. Transparent mode is typically used for voice: the input service data units (SDUs) are the packets produced by the AMR codec, and are mapped directly onto the output protocol data units (PDUs).

*Unacknowledged mode* (UM) is suitable for real time packet data such as streaming video. The input SDUs are typically IP packets, which are too large to transmit as they are. The RLC therefore cuts them into smaller PDUs, and can splice them together to make a PDU containing the end of one SDU and the beginning of the next one (Figure 3.16). It also adds a

Table 3.1 *Example of a transport format combination set. (a) Transport format sets for two simple transport channels. Each transport format is labelled using a transport format indicator (TFI). (b) Transport format combination set that supports six of the nine possible combinations from the two channels. Each combination is labelled using a transport format combination indicator (TFCI).*

(a) Transport format sets

| Channel | TFI | Block | # blocks |
|---|---|---|---|
| 1 | 0 | 0 | 1 |
| | 1 | 100 | 1 |
| | 2 | 200 | 1 |
| 2 | 0 | 100 | 0 |
| | 1 | 100 | 1 |
| | 2 | 100 | 2 |

(b) Transport format combination set

| TFCI | TFI for channel 1 | TFI for channel 2 |
|---|---|---|
| 0 | 0 | 0 |
| 1 | 0 | 1 |
| 2 | 0 | 2 |
| 3 | 1 | 0 |
| 4 | 1 | 1 |
| 5 | 2 | 0 |

header that includes a *sequence number* (SN), so the receiver can notice if any PDUs are missing.

The most complex RLC mode is *acknowledged mode* (AM). This is the only mode that does retransmissions, so it is used for non-real time packet data. It includes all the features of unacknowledged mode, together with

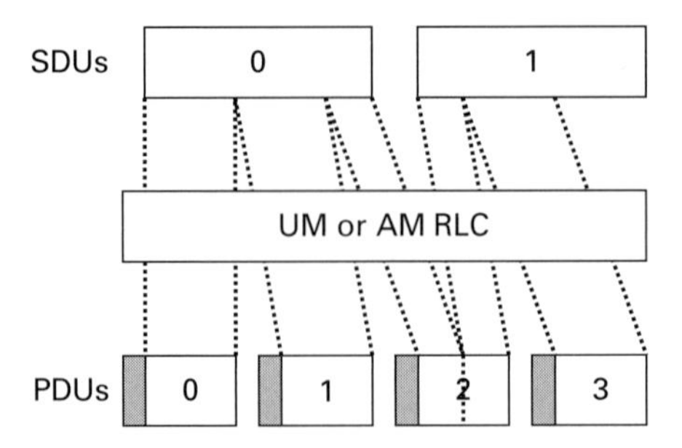


**Figure 3.16** Segmentation and concatenation in the unacknowledged and acknowledged mode RLC.

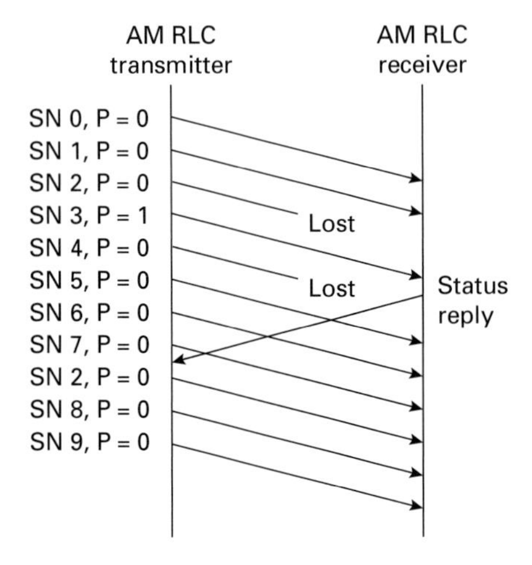


**Figure 3.17** Operation of selective retransmission in the acknowledged mode RLC.

the selective retransmission scheme illustrated in Figure 3.17. The transmitter streams packets to the receiver, and also stores them in a buffer in case they need to be retransmitted. Occasionally, it sets a polling bit P in the packet header to request acknowledgements. The receiver replies with status messages that indicate which sequence numbers have passed the cyclic redundancy check and which have failed. The transmitter can then discard the packets that arrived correctly, and retransmit the ones that did not.

When operating in unacknowledged and acknowledged mode, the RLC also encrypts the transmitted data. A quirk of the specifications is that, in transparent mode, encryption is delegated to the MAC.

## 3.2 High speed packet access

The phrase *high speed packet access* (HSPA) describes a collection of techniques that were introduced in releases 5 to 7. HSPA increases the throughput of the air interface, but it makes the system more complex, and it makes the data rate for each individual mobile more variable. This trade-off means that HSPA is only suitable for streams that can tolerate a variable bit rate, so it is only used for packet switched data streams, not for circuit switched ones.

The techniques were introduced in three stages. *High speed downlink packet access* (HSDPA) was the first implementation to appear, in release 5. Release 6 introduced the *enhanced uplink*, which is more often known as *high speed uplink packet access* (HSUPA). Release 7 brought a number of enhancements, which are collectively known as *HSPA evolution* or *HSPA+*.

The approach we will take is to describe the underlying techniques, and then show how those techniques are used in each of the three implementations. Although HSUPA was introduced second, it is actually closer to the release 99 specifications than HSDPA is. We will therefore cover the implementations in the order HSUPA, HSDPA and finally HSPA+.

### 3.2.1 Hybrid ARQ with soft combining

The first technique we will cover is *hybrid automatic repeat request* (hybrid ARQ) *with soft combining*. This is used in a similar way in both HSUPA and HSDPA, and mainly affects the bit rate processing in the physical layer.

Hybrid ARQ is the process of correcting errors by the combination of error correction and retransmissions, as described in Sections 3.1.6 and 3.1.8 above. The rationale for adding soft combining to this process comes from thinking about the acknowledged mode RLC. If the receiver's RLC picks up a transport block that has failed the cyclic redundancy check, then it discards it and asks the transmitter for a new one. This works fine, but there is a drawback: the first transport block had some useful signal energy, which has just been lost. If we could find a way to keep the first

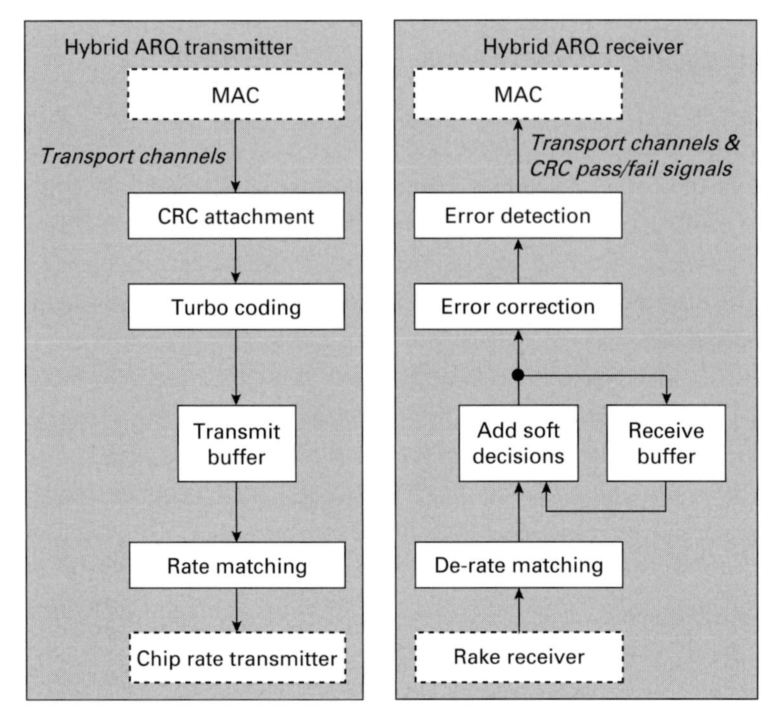


**Figure 3.18** Modifications to the bit rate processor when using hybrid ARQ with soft combining.

transport block and combine it with the second one, then we would do better than by using the second one alone.

When using hybrid ARQ together with soft combining, we achieve this by introducing an extra retransmission stage into the physical layer (Figure 3.18). The bit rate transmitter has a buffer after the error correction coder, in which it stores the coded bits in case a retransmission is required. It always uses turbo coding, because high speed packet access is aimed at the high bit rates for which turbo coding is more suitable.

In the receiver, the soft decisions are processed by the lower parts of the physical layer and are stored in a receive buffer, before passing through error correction and error detection as before. Depending on the result of the CRC, the receiver sends a positive or negative acknowledgement back to the transmitter, using a new physical layer signalling message. After a negative acknowledgement, the transmitter sends the transport block again, and the receiver combines the soft decisions from the two transmissions by adding them together. This increases the signal-to-interference ratio, so it increases the probability of the block passing the CRC. (Note that the soft decisions must be added before we reach the error correction stage, as this process is non-linear.)

The process can continue for several retransmissions until eventually we get a CRC pass. One potential problem is that the information in the receive buffer may get corrupted, perhaps by a burst of interference in one of the transmissions. To handle this possibility, the transmitter can give up and move on to the next transport block if it reaches some maximum number of retransmissions. If this happens, then the receiver's RLC notices the CRC failure, and handles the problem in the same way as in release 99.

At a system level, the benefit of this process is that it can use a higher block error ratio than the release 99 scheme, so it can use a lower transmitted signal power. As we will see in Section 3.3, this reduces the interference level in the cell, and increases its capacity. Clearly, however, it requires a data stream that can tolerate the jitter arising from retransmissions.

There is some extra subtlety buried in the rate matching stage. For a retransmission, the transmitter can use either the same puncturing or repetition pattern that it did originally, or a different one. The two cases are sometimes known as single or multiple *redundancy versions*. The processing in the two cases is nearly the same: the only differences are that the receiver needs more storage to handle multiple redundancy versions, and may combine soft decisions that are zero in one transmission and non-zero in another.

Another subtlety comes from the way in which acknowledgements and retransmissions are scheduled. HSPA uses a simpler retransmission scheme than the RLC, known as *stop-and-wait*. Using this technique, the transmitter stops after sending a transport block and waits until it has received an acknowledgement, before deciding whether to retransmit the old block or pick a new one. This works fine, but it introduces pauses into the transmitted data stream that greatly reduce the data rate.

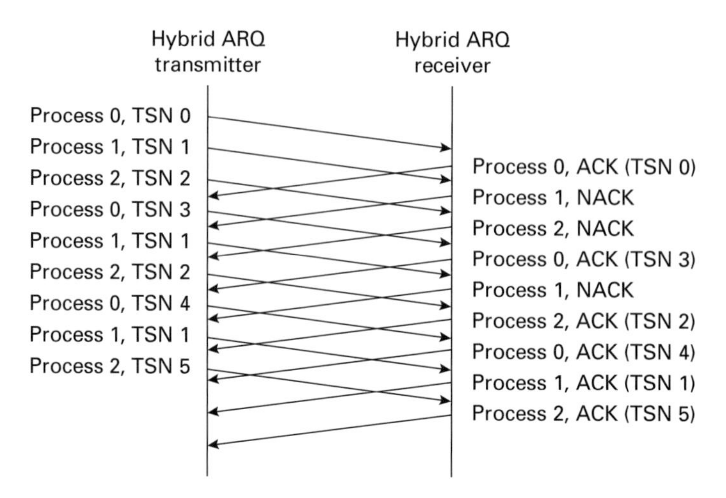


**Figure 3.19** Operation of stop-and-wait retransmission when using three hybrid ARQ processes.

The solution is to use multiple parallel versions of Figure 3.18, each of which is known as a *hybrid ARQ process*. Every transmission time interval, the transmitter sends a block on one of the hybrid ARQ processes, while waiting for acknowledgements on the others. This allows the transmitter to send data in every transmission time interval, which maximises the data rate on the air interface. The effect, for the case of three processes, is shown in Figure 3.19. In the figure, positive acknowledgements are denoted ACK, while negative acknowledgements are denoted NACK.

The only problem is that different blocks may require different numbers of retransmissions, so they may emerge from the receiver's physical layer in a different order from the one expected. To handle this, the transmitter labels each block with a transmit sequence number, denoted TSN. In the receiver, the medium access control protocol inspects the transmit sequence number, and uses it to put the blocks back in the right order.

### 3.2.2 Fast scheduling

A second technique used for high speed packet access is fast scheduling of transmissions by the Node B. The technique is implemented a bit differently in HSUPA and HSDPA, so we will describe uplink fast scheduling here, and downlink fast scheduling later on as part of Section 3.2.4.

When a mobile is sending packet data, its data rate can be highly variable. The network therefore has to limit the rate at which each mobile transmits, to minimise the likelihood of a cell reaching its maximum data rate. In release 99, this is done by means of RRC signalling messages that are exchanged between the mobile and its serving RNC. Typically, the mobile indicates that it has data to transmit, and the SRNC responds by increasing its maximum permitted data rate. Unfortunately the signalling messages take a long time (typically hundreds of milliseconds), which makes the scheduling process inefficient. When a mobile reaches the end of its data stream, for example, there is a significant delay before its resources can be allocated to another one.

We can speed up the scheduling process by moving it to the Node B. The mobile sends the Node B a signalling message, to indicate the rate at which it would like to transmit. The Node B examines the mobile's request in the light of the current load in the cell, and reacts with a signalling message of its own, to indicate the maximum data rate at which the mobile will be allowed to transmit. The mobile transmits its data, and the Node B signals its acknowledgements as described above. This control loop does not involve the RNC, so it works much faster than the equivalent loop in release 99, and is much more responsive to changes in the network load.

If the cell is lightly loaded, then the scheduling process is easy: the Node B can just allow each mobile to transmit at the data rate that it requests. If the cell is congested, then the process is harder. At one extreme, the Node B can allocate the same maximum data rate to every mobile in the cell. At the other extreme, the Node B can maximise a cell's throughput by allocating a high data rate to the mobiles that can handle it, and a low data rate, or even zero, to the others. The usual strategy lies somewhere between these two extremes, so the scheduling algorithms are not defined by the standards; instead, they are proprietary to the equipment manufacturer or the network operator.

### 3.2.3 HSUPA

HSUPA was introduced in release 6. It combines the processes of fast scheduling and hybrid ARQ with soft combining, to increase the rate at which packet data can be transmitted on the uplink.

When a mobile is using HSUPA, it sends packet data on a new transport channel, called the *enhanced dedicated channel* (E-DCH). At the same time, the mobile continues to transmit on a release 99 dedicated channel (DCH), which handles power control and any circuit switched communications. The E-DCH uses a transmission time interval of either 10 ms or 2 ms: this is less than in release 99, which speeds up the scheduling process. It supports soft handover, but the E-DCH's active set can be a subset of the DCH's active set: a cell can be in the second of these but not in the first, to support cells that do not yet implement HSUPA. One cell in the E-DCH active set is nominated as the *serving cell.*

There are five new physical channels. The network uses the *E-DCH absolute grant channel* (E-AGCH) and the *E-DCH relative grant channel* (E-RGCH) to indicate the maximum data rate at which each mobile can transmit. More precisely, the serving cell indicates an absolute data rate on the E-AGCH, and can also adjust an earlier grant up or down using shorter messages on the E-RGCH. Non-serving cells only use the E-RGCH and only to reduce the rate at which a mobile can transmit. They do this when overloaded, so as to reduce the interference they receive.

The mobile transmits data on the *E-DCH dedicated physical data channel* (E-DPDCH). Depending on its capabilities, it can use up to two channelisation codes with a spreading factor of two, plus two channelisation codes with a spreading factor of four. It also sends physical layer signalling information on the *E-DCH dedicated physical control channel* (E-DPCCH). The best known signal is the *happy bit,* which is set to 1 if the mobile is happy with its current maximum data rate, or 0 if it would like to transmit at a higher data rate and has enough power to do so. The E-DPCCH also indicates the transmitted block size, and whether the block is a retransmission or a new one. Additional signalling is sent in the MAC header of the E-DPDCH, to give the Node B more information about the amount of data available for transmission.

Each Node B sends an acknowledgement on the *E-DCH hybrid ARQ indicator channel* (E-HICH). In the mobile, a hybrid ARQ process can move on to a new block if it receives a positive acknowledgement from just one Node B.

The medium access control protocol is more complex than in release 99. In the network, there are two new components. The MAC-e is implemented in the Node B. It schedules the mobiles' uplink transmissions, in response to their uplink signalling messages and the current load in the cell. The MAC-es is implemented in the serving RNC. It receives blocks of positively acknowledged data from the Node Bs, and puts them back in their original order. In the mobile, there is a single new component, known as the MAC-e/es.

### 3.2.4 Adaptive modulation and coding

The next technique we need to consider is *adaptive modulation and coding*. This is only used in HSDPA, and arises from a difference in the way transmit power is allocated in the uplink and downlink: in the uplink each mobile can work out its transmit power independently, but in the downlink the cell has to share its power among all the mobiles that it is transmitting to. That begs the following question: how does the cell allocate its transmit power in the best possible way? In particular, how should the cell respond if a mobile moves into a fade?

We have already seen the technique used in release 99: fast power control. If a mobile moves into a fade, then the Node B increases the power transmitted to it, so as to keep the received signal-to-interference ratio constant. This in turn allows the transmitted data rate to stay the same, in accordance with Equation (1.1). For real time, constant bit rate services such as voice, this is ideal. However, the extra transmit power is unavailable for the other mobiles in the cell, which might have used it more effectively.

For non-real time services, it is better to transmit to every mobile with a constant power, and to let the data rate vary. If a mobile moves into a fade (Figure 3.20), then the received signal-to-interference ratio drops, so the Node B has to reduce the data rate that it sends there. However, the

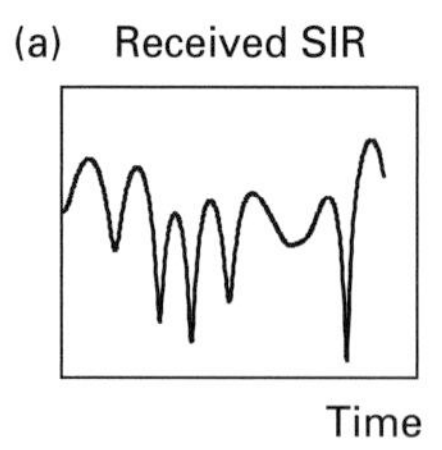


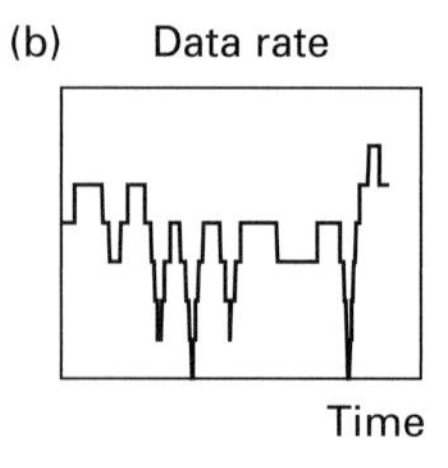


**Figure 3.20** Operation of adaptive modulation and coding in HSDPA. (a) Received signal-to-interference ratio. (b) Transmitted data rate.

extra power is then available for the other mobiles in the cell. If the Node B uses that power for mobiles that are outside a fade, it can increase the data rate that it sends to them. Crucially, the cell's total data rate is higher than it was when using fast power control.

The process works as follows. The mobile measures the received signal-to-interference ratio, and indicates the maximum data rate it can handle using a physical layer signalling message on the uplink. The Node B then varies its transmitted data rate in two ways. In adaptive coding, it transmits with a constant spreading factor, but varies the amount of puncturing or repetition in the rate matching algorithm. By doing this, it varies the number of transport channel bits per frame, and hence the data rate. In adaptive modulation, it uses not only QPSK but also a modulation scheme known as 16-QAM (*quadrature amplitude modulation*). This uses a constellation containing 16 symbols instead of four, to transmit four bits in parallel instead of two. By switching between the two modulation schemes, the Node B is able to vary the transmitted data rate further.

The effect is a fast scheduling process, in which the Node B schedules its transmissions in response to signalling messages from the mobiles. As in the uplink, the scheduling algorithm is proprietary to the equipment manufacturer or the network operator.

### 3.2.5 HSDPA

HSDPA was introduced in release 5. It combines the techniques of hybrid ARQ with soft combining, fast scheduling, and adaptive modulation and coding, to increase the rate at which packet data can be transmitted on the downlink.

When using HSDPA, the mobile receives packet data on a new transport channel, which is known as the *high speed downlink shared channel* (HS-DSCH). As in HSUPA, the mobile continues to receive information such as circuit switched communications on a release 99 DCH. One difference from HSUPA is that the transmission time interval is always 2 ms. Another is that the HSDPA channels do not use soft handover: instead, all the communications are with the same serving cell that was singled out in the description of HSUPA. The main reason is that, on the downlink, the combination of soft handover with soft combining and fast scheduling would be extremely complex.

There are three new physical channels. The Node B sends data to the mobiles on the *high speed physical downlink shared channel* (HS-PDSCH). This uses a fixed spreading factor of 16, but it can be transmitted on up to 15 channelisation codes, which can be sent either to a single mobile or to many different ones. The Node B also sends the mobiles signalling messages on the *high speed shared control channel* (HS-SCCH). These alert mobiles to the imminent arrival of data on the HS-PDSCH; they also describe the characteristics of the data, such as the channelisation codes that will be used, the modulation scheme (QPSK or 16-QAM), and whether the data will be a retransmission or a new one.

A mobile sends acknowledgements to the network on the *high speed dedicated physical control channel* (HS-DPCCH). It also uses the HS-DPCCH to send signalling information known as *channel quality indicators* (CQIs): these indicate the highest data rate that the mobile can handle, given the signal-to-interference ratio that it is currently receiving. The Node B reacts to the channel quality indicator by adjusting its transmitted data rate on the HS-PDSCH.

There is one new component to the medium access control protocol, which is known as the MAC-hs. On the network side, this is implemented in the Node B, and schedules the Node B's transmissions in response to the mobiles' channel quality indications and the current load in the cell. In the mobile, the MAC-hs receives blocks of data from the physical layer, and re-orders them.

### 3.2.6 MIMO antennas

The final technique we will describe is *multiple input multiple output* (MIMO) antennas, which have been introduced in release 7 as part of HSPA+. In Section 1.3.4, we saw how a communication system could use multiple transmit or receive antennas for diversity processing, to increase the received signal strength and reduce the amount of fading. MIMO systems use multiple antennas at both the transmitter and the receiver, but for a completely different purpose: to increase the data rate. Figure 3.21 shows a MIMO system with two antennas at the transmitter and two at the receiver, which is the greatest number that the UMTS specifications currently support.

The transmitter divides the data stream in two, and sends half the data to each antenna. The receiver is configured as a *beamforming* system, which adds together the signals that arrive at the two receive antennas so that they interfere. Signals from some directions interfere constructively while others interfere destructively, so the effect is to synthesise a receive antenna beam that has maxima in some directions (where the interference is constructive) and nulls in others (where it is destructive). By

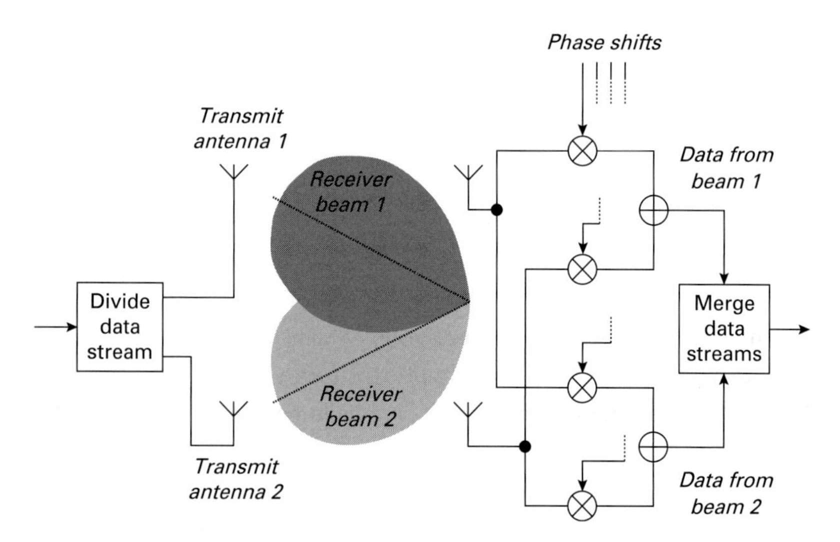


**Figure 3.21** Simplified architecture of a MIMO antenna system.

applying additional phase shifts to the two signals before they are added, we can steer the nulls to any direction we choose.

In the MIMO receiver, we use two pairs of phase shifts to synthesise two independent antenna beams. We then steer beam 1 so that it has a null in the direction of transmit antenna 2, which ensures that it mainly receives data from transmit antenna 1. Similarly, we point a null of beam 2 towards transmit antenna 1, so that it mainly receives data from transmit antenna 2. We have then set up two nearly independent data streams, from antennas 1 and 2 in the transmitter to beams 1 and 2 in the receiver, so we can immediately double the data rate. This is, at least in theory, a highly effective way to get round the limit imposed by Equation (1.1).

It will come as no surprise to learn that this explanation is greatly simplified, and there are a few difficulties in the implementation of MIMO systems. The most important is fading. This introduces random, time dependent phase shifts into the rays that travel from transmitter to receiver, which makes it harder to work out the phase shifts to apply in the receiver. It turns out that we can only work out the phase shifts if the four fading patterns, from the two transmit antennas to the two receive antennas, are independent of each other. Any correlations between them degrade the performance of the MIMO receiver, and reduce the data rate that can be achieved.

### 3.2.7 HSPA+

Release 7 introduces a number of enhancements to high speed packet access, which are collectively known as *HSPA evolution* or *HSPA+*. The most important is the introduction of MIMO antennas for HSDPA, which implements the techniques described above with a maximum of two transmit and two receive antennas. HSPA+ also supports higher order modulation schemes. The uplink can transmit four bits in parallel using 16-QAM; the downlink can transmit six bits in parallel using 64-QAM, although it is limited to 16-QAM when using MIMO antennas.

Other enhancements reduce the mobile's power consumption at the times when it is not transmitting or receiving, and allow the HS-DSCH to be used without an accompanying DCH. Finally, there are improvements to the ways in which the network handles the data, which reduce the end-to-end delay and the delay jitter. This last set of improvements is particularly beneficial, as it reduces the delay jitter to values well below 100 ms, and makes high speed packet access usable for real time communications using voice over IP.

## 3.3 Performance of UMTS

The technologies described in Sections 3.1 and 3.2 allow UMTS to support higher data rates than earlier systems. Here, we will give an indication of the data rates that can be achieved. Two numbers are particularly important: the maximum data rate that a mobile can reach under ideal conditions, and the total data rate that a cell can typically handle. The first of these can be calculated directly from the specifications, and is the figure usually quoted in marketing literature. In practice, however, the second number is usually the one that limits the performance of the system, and is the main topic of this section.

We begin by discussing how the capacity of the uplink can be estimated in release 99. After describing the differences that appear in the downlink, we show how the capacity is increased by the use of high speed packet access, and summarise the advantages and disadvantages of CDMA compared with other multiple access techniques.

### 3.3.1 Behaviour of the CDMA uplink

The behaviour of the CDMA uplink is illustrated in Figure 3.22. The base station is trying to process the signal that it is receiving from the mobile on the left, against a background of interference received from all the other nearby mobiles. The interference arises because the other mobiles are using different scrambling codes from the first one, so their transmissions are uncorrelated but not orthogonal.

If there are only a few mobiles, then the interference levels are low, and the signal reception conditions are good. As the number of mobiles increases, so the interference levels increase. We can quantify the increase in interference using a number called the *noise rise*, NR, which

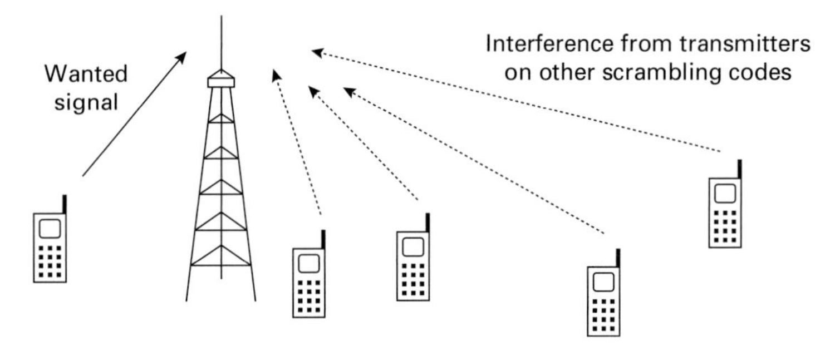


**Figure 3.22** Origin of uplink interference in CDMA.

is defined as follows:

$$NR = \frac{P_{\text{total}}}{P_{\text{thermal}}} \tag{3.2}$$

where $P_{\text{thermal}}$ is the power at the base station receiver due to thermal noise, and $P_{\text{total}}$ is the total power received from all sources, essentially thermal noise plus interference from all the UMTS mobiles.

In the uplink, the noise rise is relatively easy to estimate: some example calculations are in reference [1]. Figure 3.23a is a typical result, which shows how the noise rise varies with the total data rate in the cell, and hence with the total number of mobiles. As the total data rate increases, so the noise rise increases, first linearly and then increasingly quickly. Eventually, the noise rise becomes infinite at a data rate known as the *pole capacity*.

We can understand what is happening as follows. If the number of mobiles in the cell increases, so the interference levels at the Node B rise. The Node B can overcome the interference using the power control algorithm, by telling each mobile to increase its transmit power. Unfortunately this increases the interference further, and a vicious circle develops: each rise in interference requires more transmit power, which in turn leads to more interference. Eventually, at the pole capacity, the interference levels are so high that the base station can no longer hear any of the mobiles in the cell, no matter how powerfully they transmit. Thus the pole capacity is an absolute limit on the capacity of the cell.

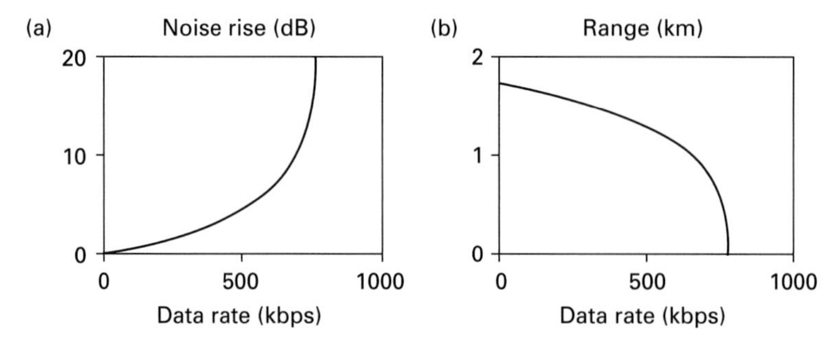


**Figure 3.23** Behaviour of the UMTS uplink. (a) Example of how the noise rise varies with the cell throughput. (b) Example of how the uplink range varies with the cell throughput.

Interference affects the cell's range as well, as shown in Figure 3.23b. As the interference levels and the noise rise increase, so the allowed propagation loss falls, and the uplink range falls. By the time we reach the pole capacity, the range has dropped to zero: the base station cannot hear any of the mobiles, no matter how close they are. The variation of range with throughput in CDMA is often known as *cell breathing*. Cell breathing is not observed in systems that use FDMA or TDMA, where the cell's capacity is determined by the number of frequencies or timeslots, and is more-or-less independent of its range.

It is important to note that the pole capacity is not a fixed quantity. Instead, the uplink pole capacity can be estimated as follows:

$$T_{\max} \approx \frac{W}{(1+f)(E_b/N_0)} \tag{3.3}$$

where $T_{\max}$ is the uplink pole capacity and $W$ is the chip rate. The number $f$ depends on the cells' geometry: it is small if the cells are isolated, and large if they overlap and interfere with each other. $E_b/N_0$ is the minimum signal-to-noise ratio per bit required for satisfactory reception: it is low if the received signal power is constant and high if there is a lot of fading, and also varies a little with the target block error ratio and the underlying bit rate. In Figure 3.23, $f$ is 0.6 and $E_b/N_0$ is 5 dB (i.e. a ratio of about 3.2), so the pole capacity in this example is about 760 kbps. In release 99, the pole capacity can vary between about 500 and 2000 kbps.

For a given pole capacity, the maximum number of mobiles in the cell depends on their average bit rate. If the pole capacity is 760 kbps, for example, then the cell can handle 62 mobiles at a rate of 12.2 kbps each, but only 11 mobiles at a rate of 64 kbps. This happens because a higher bit rate requires a higher transmit power to get the same energy into each bit, and this increases the interference that the base station receives. It is therefore better to think of the cell's capacity in terms of the total data rate it can handle, rather than the total number of mobiles in the cell.

The pole capacity is not a practical limit, because by the time the data rate reaches the pole capacity, the uplink range has dropped to zero. To deal with this problem, a cell normally operates using a *noise rise limit.* This is the maximum noise rise permitted within the cell, and leads to a practical capacity that is rather less than the pole capacity. If the noise rise starts to approach the limit, then the radio network controller has several ways of dealing with the situation. For example, it can reduce the bit rates of voice calls, delay the transmission of non-real time packet data, or block users when they try to make new calls.

In Section 3.1.3, we stated that the highest capability mobiles could transmit at a data rate of about 2048 kbps. This is usually greater than the capacity of the cell. A mobile can only reach such high data rates under ideal conditions: it must be the only mobile in the cell that is transmitting, it must be close to the base station to prevent it from hitting its maximum transmit power, and the pole capacity must be unusually high. These conditions can sometimes be met in picocells and even microcells, but are almost impossible to achieve in macrocells. As a result, it can be very difficult for a high capability mobile to reach its maximum data rate.

### 3.3.2 Behaviour of the CDMA downlink

The CDMA downlink is harder to model than the uplink, but the usual treatment runs as follows. We begin by assuming that the cell's range is limited by the uplink. This assumption is generally correct, because a base station can generally afford a powerful enough transmitter to handle all the mobiles in the cell. (In UMTS, most mobiles have a maximum power of 21 dBm ($\frac{1}{8}$ watt), while a macrocell's power is around 43 dBm (20 watts).)

We then estimate the downlink power required to reach each mobile, taking into account the interference from other base stations that are transmitting on different scrambling codes. By adding the individual power requirements, we can estimate the total power that the base station needs to transmit. If this is less than the power available at the base station, then the downlink will operate satisfactorily.

A typical result is shown in Figure 3.24, for conditions that are similar to the ones used earlier. As we saw on the uplink, the transmit power increases first linearly, and then increasingly quickly as interference comes to dominate. In this example, we reach the downlink pole capacity at a throughput of about 690 kbps, although we cannot quite get there because of the limit imposed by the base station's maximum power.

The downlink pole capacity is governed by a slightly different equation and, depending on the circumstances, can be greater or less than that of the uplink. However, the data rates on the downlink are usually greater than those on the uplink, because mobile data subscribers are likely to download far more data than they upload. As a result, we usually reach the downlink pole capacity more quickly. We can therefore summarise the behaviour of a UMTS cell as follows: coverage is usually limited by the uplink, but capacity is usually limited by the downlink.

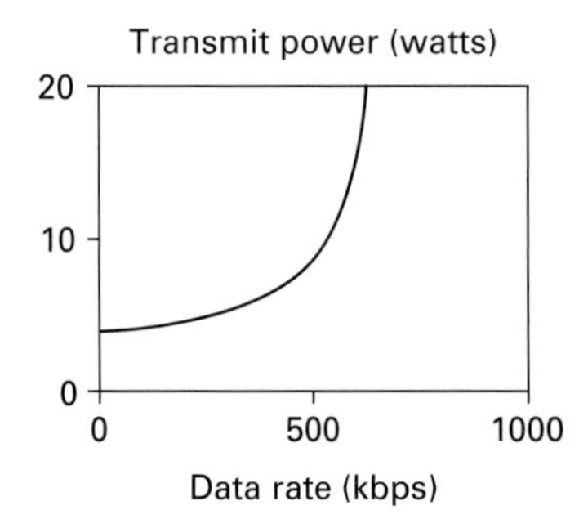


**Figure 3.24** Example of how the transmitted power varies with the cell throughput on the UMTS downlink.

### 3.3.3 Performance of HSPA

High speed packet access increases both the maximum data rate of each mobile, and the pole capacity of each cell. However, the increase in the first is usually larger.

In HSUPA, a mobile has a maximum data rate of 5.7 Mbps. This is roughly three times greater than in release 99, but can only be achieved by configuring the rate matching algorithm to use a large amount of puncturing. In HSDPA, the maximum data rate is 14 Mbps. This is roughly six times greater than before, and is achieved by using heavy puncturing in conjunction with 16-QAM. In both cases, there are only a few extra bits available for error correction. This implies that a mobile can only reach these high data rates if it lies very close to the base station, so that the received signal power is very high.

High speed packet access increases the pole capacity as well. On both the uplink and the downlink, hybrid ARQ and soft combining allow the system to be operated at a higher block error ratio than in release 99. This reduces the transmitted signal power, so it reduces the amount of interference in the cell, and increases its pole capacity. On the downlink, adaptive modulation and coding allows the base station to direct its transmit power to the mobiles that can support a high data rate, and this increases the downlink capacity further.

The increase in pole capacity is hard to estimate, as it depends on the exact conditions being used. As an example, however, reference [2] suggests that HSUPA can increase a cell's capacity by about 30 to 90 per cent, and HSDPA can increase it by a factor of about 2 to 3. While significant, this is less than the increase in the peak data rate that we noted above. As a result, it is even harder for a high capability mobile to reach its maximum data rate when using high speed packet access than it was in release 99.

### 3.3.4 Advantages and disadvantages of W-CDMA

We close this chapter by summarising the main advantages and disadvantages of wideband CDMA, when compared with other multiple access techniques. There are three main advantages.

The first advantage comes from the fact that a cell's capacity is limited by interference. If a mobile is not transmitting, then it does not cause any interference to the base station, so it doesn't take up any uplink resources. Similarly it doesn't take up any downlink resources if it is not receiving. This is particularly useful in a voice call, where a typical mobile is only transmitting information for half the time, and receiving for the other half. The capacity improvement is not as great as it might be, because the mobile still has to transmit and receive continuously on the DPCCH. Despite this, the effect increases the cell's voice capacity by about 50 per cent.

Second, the short chip duration allows a receiver to do multipath diversity processing, using the rake receiver that we described earlier. This reduces the amount of fading, and allows the system to work with a lower received signal power than it otherwise would. In turn, this reduces the amount of interference, and increases the capacity of the system.

Third, the cells in a CDMA network are distinguished by the use of different scrambling codes, so they can all use the same carrier frequency. This is a different situation from the one in FDMA or TDMA, in which nearby cells have to use different carrier frequencies, and means that a CDMA network can accommodate more mobiles per unit bandwidth. Radio spectrum is a scarce and sometimes expensive resource, so this can be an important advantage.

CDMA does, however, bring a couple of problems. The first is related to radio network planning. Interference between neighbouring cells means that we cannot plan each cell independently, as we could in FDMA or TDMA. Instead, changes to one cell can affect the behaviour of cells nearby. If, for example, we increase the power transmitted from one cell in an attempt to increase its range, then that will increase the interference to the downlinks of neighbouring cells and degrade them. These inter-relationships make radio network planning a more complex process than it has previously been.

Another problem is related to intellectual property. CDMA technology is covered by a large number of patents, which require substantial royalties from product developers, and have been slowing the wider adoption of UMTS. Some of the licensing terms are the subject of ongoing legal action and, at the time of writing, the eventual outcome is still unclear.

## References

1. H. Holma & A. Toskala, *WCDMA for UMTS: HSPA Evolution and LTE*, 4th edition (Wiley, 2007).
2. J. Sköld, M. Lundevall, S. Parkvall & M. Sundelin, Broadband data performance of third-generation mobile systems. *Ericsson Review*, **82**:1 (2005), 14–23.

# 4 Operational procedures

In UMTS, the network elements communicate with each other by exchanging signalling messages, which are written using the signalling protocols that we introduced in Chapter 2. The signalling messages are organised into procedures, which define how the network elements interact with each other, and which ultimately control the operation of the system. These signalling procedures are the main theme of the next two chapters. In this chapter, we discuss the procedures which control the internal operation of the system, and which do not involve any communication with the outside world. In Chapter 5, we will discuss the procedures that are related to particular services, such as voice and GPRS.

Here, we start by reviewing the way in which the network manages its communications with the mobile, and the different internal states that a mobile can be in. We then describe the procedures that a mobile uses when it switches on, to discover the cells around it and establish communications with the network. This leads to a discussion of the techniques that are used to keep the system secure in the presence of intruders. The second half of the chapter describes the procedures that take place inside a mobile after it has switched on. These are covered in two sections, as the exact procedures depend on the internal state that the mobile is in. The chapter closes by describing how the mobile stops communicating with the network and switches off.

We will make several references in this chapter to the architectural components and software protocols that we introduced in Chapter 2. Before going further, the reader may find it useful to review that chapter, particularly Sections 2.2 and 2.3.

## 4.1 Management of signalling connections

During its operation, the mobile has to exchange signalling messages with three different networks: the circuit switched and packet switched domains of the core network, and the radio access network. Each set of signalling messages is managed by a *signalling connection*, which can be thought of as a communication pathway between the mobile and the corresponding network. When a signalling connection is present, the mobile and the network can freely communicate with each other; when it is absent, only a limited amount of communication is possible.

Each signalling connection is associated with a state diagram, which describes the ways in which the mobile and the network can communicate. At a simple level, we would expect each diagram to have two states, corresponding to the presence and absence of a signalling connection. In practice, we will see that the state diagrams are slightly more complex than that. It will be worth remembering the names of the states described below, as they will come up a lot in the sections that follow.

### 4.1.1 Core network

Figure 4.1 shows the state diagrams that describe the mobile's communications with the core network.

The state diagram for the circuit switched domain has three states (Figure 4.1a). In CS-CONNECTED state, the circuit switched domain has a signalling connection with the mobile. It uses this connection

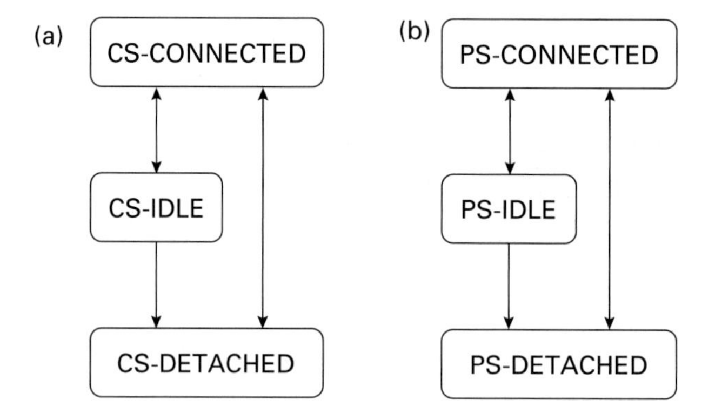


**Figure 4.1** State diagrams that describe the mobile's communications with the core network. (a) Circuit switched domain. (b) Packet switched domain. (Adapted from 3GPP TS 23.060.)

to exchange signalling messages with the mobile, in order to control services like phone calls and system related issues like security procedures. This state is typically used for mobiles that are making a call.

In CS-CONNECTED state, the visitor location register (VLR) knows which radio network controller is serving the mobile, and it delegates knowledge of the mobile's cell to that serving RNC. It does this on the basis that cells are an issue for the radio access network alone: the core network has no business knowing about them.

CS-IDLE is typically used for mobiles that are on standby. In this state, the mobile does not have a signalling connection with the circuit switched domain, but three kinds of communication are still possible. First, the VLR knows the mobile's location to an accuracy of a *location area* (LA), which is a region of the network containing a large number of cells and one or more RNCs. When a mobile enters a new location area, it sends the network a mobility management (MM) message called *location area update request*, to keep the network informed about its location. This also happens at regular intervals, typically an hour, even if the mobile stays still.

Second, a mobile that wants to start a phone call can send the network an MM message called a *connection management service request*. The network reacts by moving it into CS-CONNECTED state, which supports the full set of signalling messages. Third, if an incoming call arrives, then the network sends the mobile a *paging* message through all the RNCs in the location area. The mobile responds with a connection management service request, as if it were starting a call of its own.

Location areas are used because they reduce the number of location update messages for mobiles on standby, at the expense of increased paging when a call comes in. The optimum size for a location area is the size that minimises the total amount of signalling.

In both of these states, the network identifies the mobile using a *temporary mobile subscriber identity* (TMSI). This is similar to the international mobile subscriber identity (IMSI), but is only assigned temporarily, so it can be broadcast without compromising the user's security. The network typically changes the TMSI as part of a location area update.

CS-DETACHED is used for mobiles that cannot be reached by the circuit switched domain, typically because they are switched off or outside the coverage areas of UMTS and GSM. In this state, the mobile and VLR record the location area and TMSI that the mobile was most recently using, for use when it switches on again.

Figure 4.1b shows the state diagram for the packet switched domain. As before, mobiles in PS-CONNECTED state have a signalling connection with the packet switched domain, mobiles in PS-IDLE can send and receive a limited number of messages, and mobiles in PS-DETACHED cannot be contacted at all. However, there are a few differences in the ways these states are used.

First, the network can save resources by putting a mobile into PS-DETACHED state, even if the mobile is still switched on. It can do this because the mobile, not the network, will initiate any future packet switched communications. This would be inappropriate in the circuit switched domain, because it would prevent the mobile from receiving incoming calls.

Similarly, the network can save resources by putting the mobile into PS-IDLE state, even if the mobile is downloading data from a server in the outside world. It typically does this if there is a long pause in the data flow, perhaps if the user is looking at a web page that has just been downloaded. To restart communications, the mobile can send a service request on the uplink, while the network can send a paging message on the downlink. This would be inappropriate in the circuit switched domain, where the data streams are continuous, not bursty.

In PS-IDLE state, the serving GPRS support node (SGSN) knows the mobile's location to an accuracy of a *routing area* (RA). Routing areas are smaller than location areas, in that each location area is divided into one or more routing areas. The reason is that we expect more paging messages in the packet switched domain, so their optimum size is less.

Finally, the packet switched domain identifies mobiles using the *packet TMSI* (P-TMSI), in place of the TMSI.

### 4.1.2 Radio access network

Figure 4.2 shows the state diagram that describes the mobile's communications with the radio access network. The state diagram is

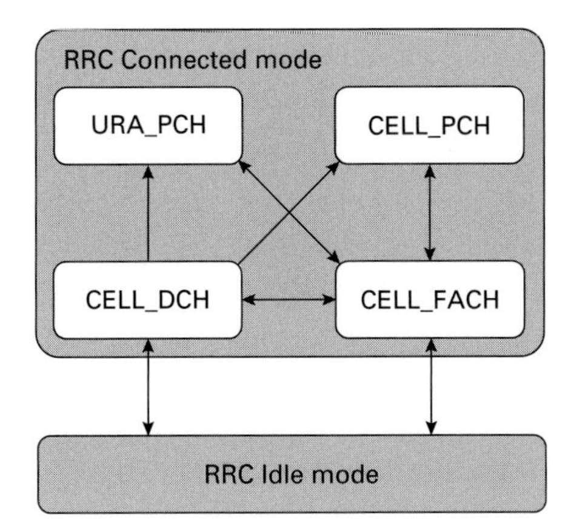


**Figure 4.2** State diagram that describes the mobile's communications with the radio access network. (Adapted from 3GPP TS 25.331.)

managed by the radio resource control protocol, which is implemented in the mobile and its serving RNC. There are two main states, known as *RRC modes*: RRC Connected and RRC Idle.

In RRC Connected mode, the mobile is assigned a serving RNC, and communicates with it using a signalling connection that is known as an *RRC connection*. Usually, but not always, the SRNC knows which cell the mobile is in.

In this mode, the network identifies the mobile using a *radio network temporary identity* (RNTI). This is similar to a TMSI but is used by the radio access network alone. There are two main types of RNTI: the *UTRAN RNTI* (U-RNTI) uniquely identifies the mobile within the operator's radio access network, while the shorter *cell RNTI* (C-RNTI) is only unique within a particular cell.

In RRC Idle mode, the mobile does not have an RRC connection. Because of this, it does not have a serving RNC or an RNTI, and the network does not know which cell the mobile is in. Even so, two kinds of communication are still possible. First, a mobile can contact the radio access network by sending it a message called an *RRC connection request*. The network reacts by moving it into RRC Connected mode and assigning it to a serving RNC. Second, the network can contact the mobile by sending it a paging message, to which the mobile responds with an RRC connection request of its own. We can see that RRC Idle mode has some similarities with the core network idle states, and others with core network detached.

As shown in the figure, RRC Connected mode is divided into four states, which depend on the network's knowledge of the mobile's location and the type of communication between them.

In CELL_DCH state, the network knows which cell the mobile is in. It communicates with the mobile using the dedicated channel (DCH), with the information flow shown in Figure 2.17a. The normal uses are for voice calls and fast packet switched data transfers.

In CELL_FACH state, the network and mobile communicate using common transport channels, namely the random access channel (RACH) and the forward access channel (FACH) (Figures 2.17b and 2.17c). This state is suitable for signalling messages and small amounts of packet data, which do not justify the overhead of a dedicated channelisation code. It might be used for an interactive game of poker, for example.

The other two states are not used so often. In CELL_PCH state, the network still knows which cell the mobile is in, but promises to send it a paging message if any data arrive on the downlink. This allows the mobile to go into a low power state known as *sleep mode* (Section 4.4.3), which increases its battery life. It might be used for the pauses between moves in an interactive game of chess. (If the game is particularly slow, then the PS domain can place the mobile into PS-IDLE, so that it uses fewer resources in the core network.)

URA_PCH is similar, but the network only knows the mobile's location to an accuracy of a *UTRAN registration area* (URA). URAs are similar to location and routing areas, but they are only used by the radio access network and can overlap, so that a cell can belong to more than one URA. In this state, the mobile only updates its location if it crosses a URA boundary. This reduces the number of location updates, so is suitable for a fast-moving mobile that would otherwise be in CELL_PCH. An obvious (if unusual) example is interactive chess on a high speed train.

Table 4.1 *Relationships between the state diagrams for the radio access network and the core network.*

<table>
<tr><th></th><th>CS-DETACHED</th><th>CS-IDLE</th><th>CS-CONNECTED</th></tr>
<tr><td>PS-DETACHED</td><td colspan="2" rowspan="2">RRC Idle</td><td rowspan="3">RRC Connected (CELL_FACH or CELL_DCH)</td></tr>
<tr><td>PS-IDLE</td></tr>
<tr><td>PS-CONNECTED</td><td colspan="2">RRC Connected (any state)</td></tr>
</table>

### 4.1.3 Relationships between the state diagrams

There are several relationships between the state diagrams described above, which apply in all except transient conditions. These relationships are summarised in Table 4.1.

If a mobile is in PS-CONNECTED state, then it is exchanging signalling messages with the core network's packet switched domain. As we saw in Chapter 2, these messages are transported using direct transfers. To handle the direct transfers, we require an RRC signalling connection between the mobile and a serving RNC, so the mobile is in RRC Connected mode. The same thing applies in CS-CONNECTED state, except that the only supported RRC states are CELL_DCH and CELL_FACH.

In PS-CONNECTED state, we also require a signalling connection on the Iu-PS interface, which transports signalling messages between the SRNC and the SGSN using RANAP direct transfers. This works in a similar way to the RRC connection, although it has no equivalent to the internal states such as CELL_DCH. The same thing applies in CS-CONNECTED state, on the Iu-CS interface.

Figure 4.3 summarises the relationships between location areas, routing areas and URAs. We can see that location areas are subdivided into routing areas, but have no relationship with URAs at all.

## 4.2 Power-on procedures

When a mobile switches on, it has to go through several different procedures before it can start a phone call or a data transfer. These procedures can be broken down into two phases.

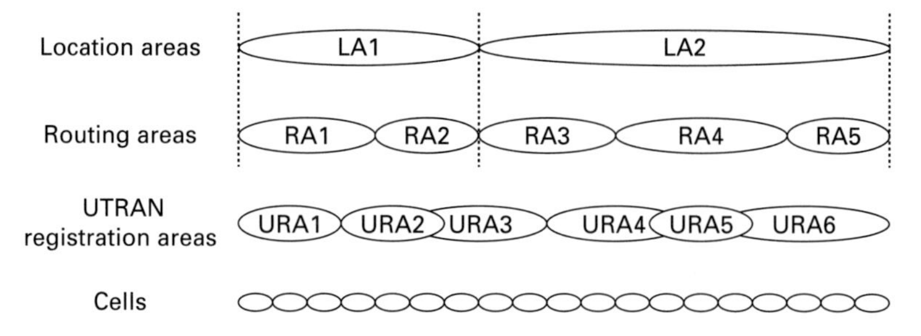


**Figure 4.3** Relationships between location areas, routing areas and UTRAN registration areas. (Adapted from 3GPP TS 23.221.)

In the first phase, the mobile discovers which cells are around it, and selects a cell that it will try to communicate with. We will give an overview of this phase in Section 4.2.1, with more detail about the underlying processes in Sections 4.2.2 to 4.2.4. At the end of this phase, the mobile has entered RRC Idle mode, and is said to be *camping* on the selected cell.

In the second phase, the mobile uses the selected cell to establish communications with the network, first with the radio access network alone (Section 4.2.5), and then with the core network (Section 4.2.6). By the end of these procedures, the mobile is fully switched on, and is ready to receive incoming calls or start a new service of its own.

## 4.2.1 Cell and network selection

The cell and network selection procedure has two modes of operation: automatic, where the choice of network is made by the mobile, and manual, where it is made by the user. The user can select one mode or other by means of the mobile's user interface.

The usual procedure is automatic mode. The procedure begins inside the mobile, when the mobile equipment (ME) asks the universal integrated circuit card (UICC) for the public land mobile network identity (PLMN-ID) of the network it was using last time it was switched on. (This network is known as the *registered PLMN*.) The mobile also retrieves information about the cells it was most recently using, such as their carrier frequencies.

The mobile then goes to its most recent carrier frequency, and runs three procedures. In the cell search procedure (Section 4.2.2), it identifies the cells that are transmitting on that frequency, together with information such as their scrambling codes. It then reads the system information that the cells are broadcasting (Section 4.2.3), to discover the PLMN-ID of the network that is using that frequency. If the PLMN-ID is the same as the one requested, then the mobile runs the cell selection procedure (Section 4.2.4). This looks for a cell whose received signal is strong enough for the mobile to use, known as a *suitable cell.* If the procedures are successful, then the mobile uses the selected cell to establish contact with the network, as we will describe in Sections 4.2.5 and 4.2.6.

If the mobile cannot find the registered PLMN on any of the stored carrier frequencies, then it examines all the other carrier frequencies that are used by UMTS, saving information about the cells it finds in case it needs that information later on. If it still cannot find the registered PLMN, then it looks for any other networks that are stored in the UICC: first the home PLMN, then any networks that the operator has defined, and finally any networks that have been listed by the user. If it is still unsuccessful, then the mobile tries to select any cell at all, regardless of the identity of the PLMN. In this final state, it can make emergency calls but nothing else.

If the mobile supports other radio access technologies such as GSM, then each PLMN-ID stored on the UICC can be associated with one or more technologies in a defined priority order. The mobile examines each of these technologies as well as UMTS, using their equivalent procedures for cell search and cell selection.

The procedure in manual mode is broadly similar, except that the mobile presents the user with a list of networks that it can see, and the user then makes the selection.

### 4.2.2 Cell search procedure

The *cell search* procedure finds the cells that the mobile can hear on a particular carrier frequency, and identifies their scrambling codes. It also measures information about the rays that the mobile is receiving from each cell, such as their amplitudes, phases and arrival times. The procedure takes place in three stages, in which the mobile examines three physical channels in turn: the *primary synchronisation channel* (P-SCH), the *secondary synchronisation channel* (S-SCH) and the *common pilot channel* (CPICH).

We introduced the synchronisation channel in Chapter 2, but it is actually divided into two components, primary and secondary. In the first step, the mobile examines the P-SCH. Unlike most of the other physical channels, the P-SCH contains a series of bursts that occupies the first 10 per cent (256 chips) of every slot. Furthermore, these bursts are not scrambled in the usual way: instead, they contain a fixed chip sequence that (at least for our purposes) is the same for every UMTS cell in the world. The mobile compares its input signal with this chip sequence, for example by passing it through a suitable matched filter. By doing this, it discovers how many rays it is receiving and, for each ray, the amplitude, the phase, and the time when a new slot begins.

The mobile now has to find the scrambling code that each ray is using. Before explaining how it does this, we need to describe how the downlink scrambling codes are organised. As shown in Figure 4.4, the downlink uses a total of 8192 scrambling codes. These are organised into 512 *scrambling code sets*, each of which contains one *primary scrambling code* and 15 *secondary scrambling codes*. Each cell uses one scrambling code set, but it nearly always transmits on the primary scrambling code alone: secondary scrambling codes are only used in

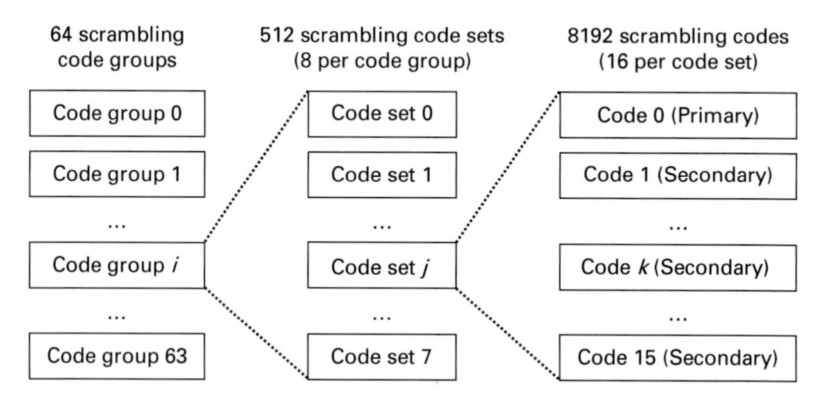


**Figure 4.4** Organisation of the scrambling codes into code sets and code groups.

unusual circumstances, and we will not consider them further. In turn, the sets are organised into 64 *scrambling code groups*, each of which contains 8 scrambling code sets.

In the second step of the cell search procedure, the mobile examines the S-SCH. This is similar to the P-SCH, in that it is transmitted in the first 256 chips of every slot. However, the chip sequence is no longer fixed: instead, it depends on the slot number within the frame (0 to 14) and the scrambling code group that the cell belongs to. By measuring the S-SCH chip sequence for each of the 15 slots in a frame, the mobile discovers the time when a new frame begins, and the scrambling code group.

In the third step, the mobile examines the CPICH. This contains a fixed bit sequence that is scrambled using the cell's primary scrambling code. (Strictly speaking, we should refer to this channel as the primary CPICH, leaving the possibility of secondary CPICHs that use secondary scrambling codes.) The mobile processes the CPICH using each of the eight possible primary scrambling codes, and the correct choice is the one that recovers the original bit sequence. The mobile now knows the primary scrambling code of each ray, so it can organise the rays into cells by assuming that rays with the same scrambling code come from the same cell.

By the end of this process, the mobile knows the primary scrambling code of each cell, and the amplitude, phase and arrival time of each ray. This is enough information to initialise the rake receiver, and to begin reading data from the cell. The mobile will continue to run the cell search procedure throughout its operation, so as to discover new rays as they appear and send information about them to the rake receiver. These tasks are carried out by the searcher function from Section 3.1.5.

### 4.2.3 System information broadcasts

*System information messages* are a special type of RRC message, which tell the mobiles in a cell about how the cell is configured. The radio access network broadcasts these messages continuously on the *broadcast channel* (BCH), which is mapped to the *primary common control physical channel* (PCCPCH) using the information flow in Figure 2.17e. The

Table 4.2 *List of the most important system information blocks, with some examples of the parameters that they contain.*

| SIB | Description | Example parameters |
|---|---|---|
| SIB 1 | Core network parameters | Location area identity<br>Routing area identity<br>PLMN-ID |
| SIB 2 | URA identities | URA identity |
| SIB 3, 4 | Cell selection and reselection parameters | $Q_{\text{qualmin}}$<br>$Q_{\text{rxlevmin}}$<br>UE_TXPWR_MAX_RACH |
| SIB 5, 6 | Common channel definitions | SCCPCH channelisation code |
| SIB 7 | Fast-changing parameters | UL interference at the Node B |
| SIB 11, 12 | Measurement control information | Scrambling codes of neighbouring cells |

PCCPCH always uses the same channelisation code, to ensure that all the mobiles in the cell can read it.

The messages are organised into *system information blocks* (SIBs), which are numbered from 1 to 18. Table 4.2 lists the most important system information blocks, together with some of parameters that they contain. At this stage, the most important information is the PLMN-ID and the parameters used by the cell selection algorithm. By reading this information, the mobile can discover whether the cell belongs to the requested network, and can proceed to the next step in the power-on procedure, known as cell selection.

The distinction between system information blocks 3 and 4 is as follows. SIB 3 is mandatory, while SIB 4 is optional. If SIB 4 is present, then mobiles in RRC Connected mode use the information that it contains, while mobiles in RRC Idle mode use SIB 3. Otherwise, all mobiles use

SIB 3. There is a similar distinction between SIBs 5 and 6, and between SIBs 11 and 12.

The system information is repeated every 4096 frames. Each PCCPCH frame is labelled using a *system frame number* (SFN) which runs from 0 to 4095, and which is used as a timing reference for all the transmissions from a particular cell. Different cells do not have to be synchronised in any way, which avoids the need to distribute accurate timing signals within the network. Instead, their frames can begin at different times, and they can be labelled with different system frame numbers.

Within the 4096 frame repetition period, the cell may send some system information blocks very often, but may only send others once or twice. (For example, the cell might transmit the fast changing parameters in SIB 7 more often than the others.) A *master information block* (MIB) is transmitted at fixed system frame numbers, and acts as a table of contents by listing the frames in which each SIB appears.

### 4.2.4 Cell selection procedure

The aim of the *cell selection* procedure is to identify a suitable cell on the selected carrier frequency that belongs to the requested network. A suitable cell is one that satisfies two criteria. First, the received signal-to-interference ratio must be high enough:

$$Q_{\text{qualmeas}} > Q_{\text{qualmin}} \tag{4.1}$$

where $Q_{\text{qualmeas}}$ is the received signal-to-interference ratio on the CPICH and $Q_{\text{qualmin}}$ is the minimum permitted value. Second, the received signal strength must be high enough:

$$Q_{\text{rxlevmeas}} > Q_{\text{rxlevmin}} + \text{Max}\,(\text{UE\_TXPWR\_MAX\_RACH} - \text{P\_MAX},\, 0) \tag{4.2}$$

The first two terms in Equation (4.2) act in a similar way to the ones in Equation (4.1): $Q_{\text{rxlevmeas}}$ is the received power of the CPICH, and $Q_{\text{rxlevmin}}$ is the minimum permitted value. The others prevent a mobile from selecting the cell if its uplink transmit power is too weak for the Node B to hear it: P_MAX is the mobile's maximum output power, and

UE_TXPWR_MAX_RACH is the maximum power that the cell will let it transmit on the physical random access channel.

The values of $Q_{\mathrm{qualmin}}$, $Q_{\mathrm{rxlevmin}}$ and UE_TXPWR_MAX_RACH are broadcast as part of SIBs 3 and 4, and can vary from one cell to another. By choosing the right values, the network can ensure that a mobile only selects a cell if it can hear the Node B on the downlink, and can transmit a strong enough signal to the Node B on the uplink.

Note that the mobile only needs to find one cell that satisfies these criteria in order to use a particular carrier frequency. It does not have to find the best cell, although a particular implementation might in fact do so.

### 4.2.5 RRC connection setup

On completion of the cell and network selection procedure, the mobile is in RRC Idle mode and is camping on the selected cell. It now establishes a signalling connection with the radio access network using a procedure called *RRC connection setup*, and moves into RRC Connected mode in either CELL_FACH or CELL_DCH state. We have already seen the protocol stack used to exchange these messages, in Figure 2.18.

Figure 4.5 shows the procedure if the eventual state is CELL_FACH. In step 1, the mobile composes a message known as an *RRC connection request*. This includes the mobile's identity, which is normally specified using the TMSI or P-TMSI that it was last using, and which the mobile has

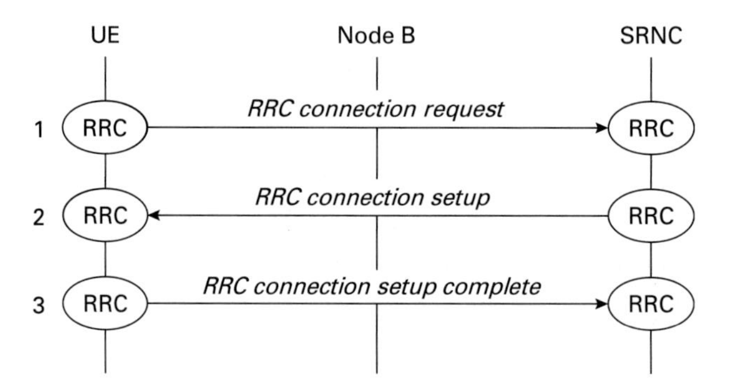


**Figure 4.5** Message sequence chart for an RRC connection setup, if the mobile's eventual state is CELL_FACH. (After 3GPP TR 25.931.)

stored in the UICC. It also includes the reason why it is sending the message, in this case a wish to establish communications with the core network, and some measurements of the signal power received from neighbouring cells.

The mobile sends the message using signalling radio bearer 0 (SRB 0), which is the radio bearer used by mobiles in RRC Idle mode. The 3GPP specifications define how SRB 0 is configured (for example, how many CRC bits it uses), so the mobile knows how to transmit it. The mobile sends the message on the random access channel (RACH), as shown in Figure 2.17b, and the Node B forwards the message to its controlling RNC.

On receiving the message, the RNC takes on the role of the mobile's serving RNC. As such, it will exchange RRC signalling messages with the mobile, and will be the mobile's sole point of contact with the core network. The RNC then composes a message known as *RRC connection setup* (2), which has several pieces of information. First, the RNC echoes back the identity that the mobile originally used, and assigns the mobile an RNTI for use in subsequent communications. It then tells the mobile which RRC state to enter: here, we assume that the RRC state is CELL_FACH, which is likely to be appropriate as the mobile is just exchanging signalling messages with the network. Finally, the RNC sends descriptions of signalling radio bearers 1 to 4, so that the mobile can use these for subsequent signalling messages.

The mobile does not yet know about the other signalling radio bearers, so the RNC sends its message using SRB 0. The message is transported using the forward access channel (FACH) and the secondary common control physical channel (SCCPCH), as shown in Figure 2.17c. The network broadcasts the SCCPCH channelisation code in SIBs 5 and 6, so the mobile knows how to receive it.

The mobile reads the message, moves to CELL_FACH state, and configures its signalling radio bearers in the manner required. Using SRB 2, it sends a confirmation message to the network known as *RRC connection setup complete* (3), and transmits it on the RACH.

The procedures in CELL_FACH state are straightforward, because the radio access network uses common channels that were configured when the cell was first set up. We need some extra signalling steps in

CELL_DCH state, to set up channels that are dedicated to the mobile in question. The process is shown in Figure 4.6.

In step 1, the mobile sends an RRC connection request as before. The RNC identifies itself as the mobile's serving RNC, but this time, it decides to put the mobile into CELL_DCH state. Using the Node B application part, it sends the Node B a message called *radio link setup request* (2), which tells the Node B how to communicate with the mobile. For example, the message includes the channelisation and scrambling codes to

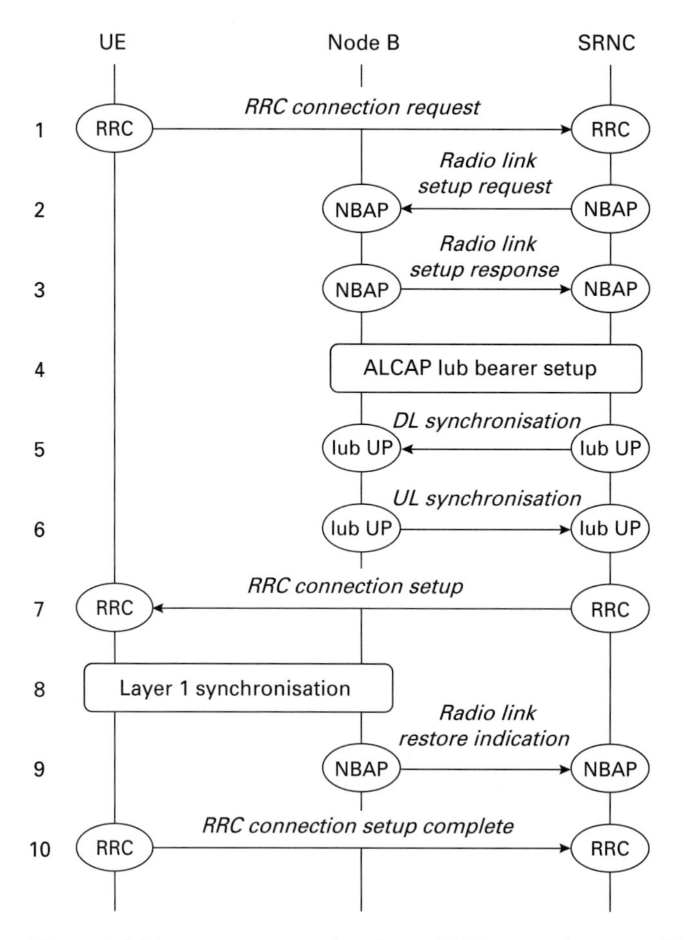


**Figure 4.6** Message sequence chart for an RRC connection setup, if the mobile's eventual state is CELL_DCH. (After 3GPP TR 25.931.)

use on the downlink, and the scrambling code that the mobile will use on the uplink. The Node B starts to listen out for the mobile, and replies with a *radio link setup response* (3).

If the radio access network is transporting data using ATM, then it needs to set up temporary virtual circuits to handle the data streams on the Iub interface. This is done by an exchange of low level messages between the SRNC and the Node B (4), using the access link control application protocol (ALCAP).

Using the Iub user plane protocol, the SRNC and Node B then exchange *synchronisation* messages (5, 6). These measure the time delays between the two network nodes so that the SRNC knows when to send information such that it reaches the Node B at the right time, and vice-versa. They also establish a common numbering scheme for the frames that are being exchanged with the mobile, which is known as the *connection frame number* (CFN). This is a similar quantity to the SFN, but it runs from 0 to 255, and is common to all the network elements that are communicating with the mobile. When the Node B has processed the downlink synchronisation message, it can start transmitting to the mobile.

If it wishes, the SRNC can send the preceding messages to more than one cell, so as to put the mobile into an immediate state of soft handover. It does this based on the signal strength measurements that the mobile sent in its RRC connection request. The cells can be controlled by the same Node B, or by different Node Bs, or by different RNCs. The last case requires some extra signalling messages on the Iur interface, which are beyond the scope of this book.

The SRNC can now compose its RRC connection setup message (7), and send it to the mobile as before. The mobile enters CELL_DCH state, and starts to transmit and receive. The Node B establishes physical layer communications with the mobile (8), and informs the SRNC (9). Finally, the mobile acknowledges the SRNC's message by sending it an RRC connection setup complete message (10). It transmits this on the dedicated channel (DCH), using the information flow in Figure 2.17a.

The mobile is now in RRC Connected mode, and has established signalling communications with the radio access network. The power-on procedure has not finished, however, as the mobile still needs to start communicating with the core network.

### 4.2.6 Core network attach

In the last stage of the power-on procedure, the mobile sets up a signalling connection with the core network. It can do this in three ways. If it can only handle circuit switched or packet switched communications, then it attaches to the CS domain using a process called an *IMSI attach*, or to the PS domain using a *GPRS attach*. Normally, however, the mobile attaches to both core network domains at the same time, using the *combined attach* shown in Figure 4.7. In this process, the mobile sends a message to the PS domain, and the PS domain sets up communications with the CS domain.

At the start of the process, we assume that the mobile is already in RRC Connected mode. In step 1, the mobile composes a GMM message called an *attach request*. This includes the P-TMSI and routing area that the mobile was last using and the type of attach that it is requesting, here a combined attach. It also indicates whether the mobile has any more signalling messages to send after this one. The mobile sends the message to its serving RNC by embedding it in an RRC *initial direct transfer*. This is the same as a normal uplink direct transfer, but it also tells the SRNC that the mobile wishes to establish a signalling connection with the core network.

The SRNC extracts the attach request, and sends it to an SGSN by embedding it in a RANAP *initial UE message*. This is the same as a normal RANAP direct transfer, but it also sets up an Iu-PS signalling connection for that mobile between the SRNC and the SGSN. On receipt of the attach request, the SGSN runs a set of security procedures (2), which authenticate the mobile and start the processes of ciphering and integrity protection. These processes are self-contained, and we will describe them in Section 4.3.

If the mobile has moved to a new routing area since it was last switched on, then the SGSN may have changed. If this has happened, then the new SGSN informs the HLR by sending it a MAP message known as *update location* (3). The HLR cancels the mobile's entry in the old SGSN (4, 5),

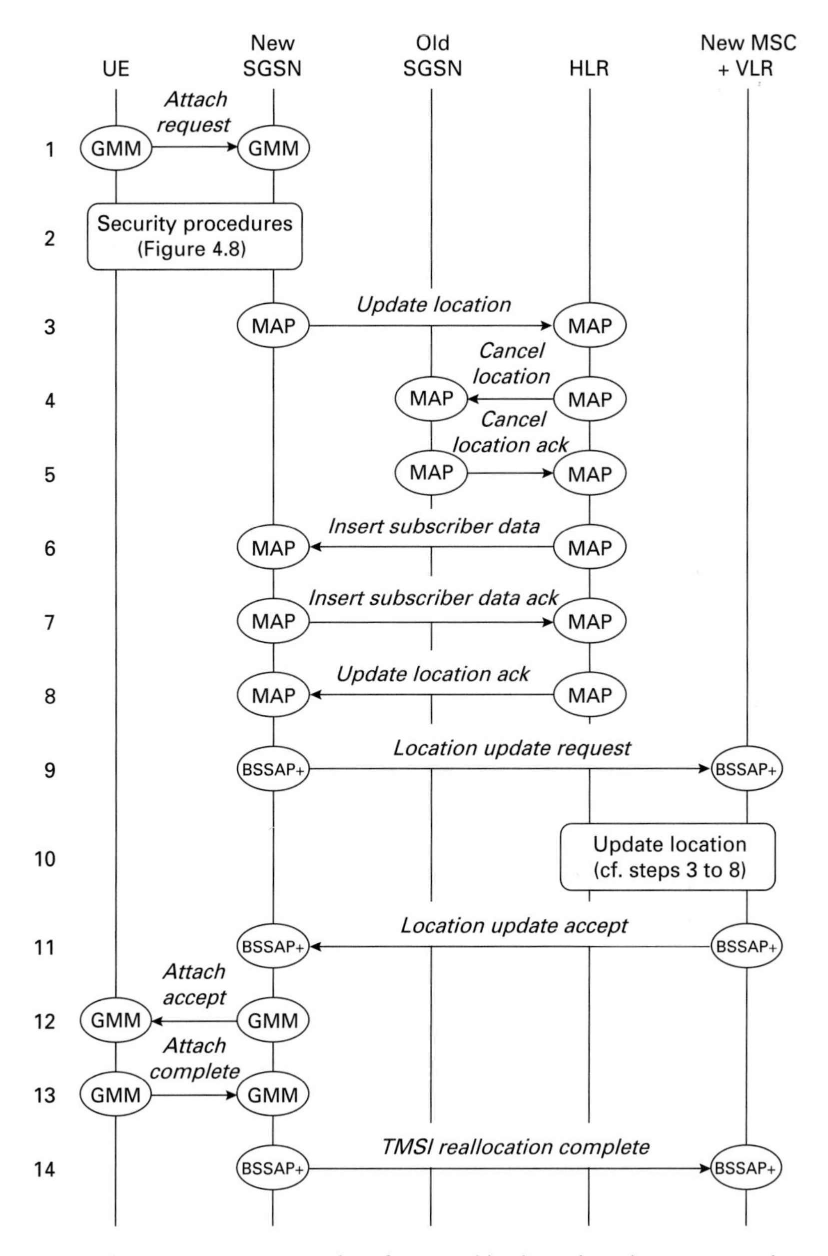


**Figure 4.7** Message sequence chart for a combined attach to the core network. (Adapted from 3GPP TS 23.060.)

sends information about the mobile to the new SGSN such as the services it is allowed to use (6, 7), and acknowledges the SGSN's original message (8). These steps are not required if the SGSN has stayed the same.

If the mobile is doing a combined attach, then the SGSN sends a BSSAP+ message called *location update request* to its corresponding MSC/VLR (9). This message requests that the mobile be attached to the CS domain as well. If the MSC/VLR has changed, then the CS domain updates the mobile's location (10) by running through its own version of steps 3 to 8. The CS domain then acknowledges the SGSN's request (11).

The SGSN can now reply to the mobile's attach request, by sending it a message called *attach accept* (12). This includes the result of the attachment process, a new P-TMSI and a new routing area identity. The mobile sends an acknowledgement to the SGSN (13), and the SGSN signals this acknowledgement to the circuit switched domain (14).

The mobile is now attached to both core network domains, but the core network states are subtly different. It is in PS-CONNECTED state, because it has set up a signalling connection across the Iu-PS interface, but in CS-IDLE state, because there is no Iu-CS signalling connection. The SGSN can move the mobile into PS-IDLE state by a further exchange of signalling messages. Alternatively, if the mobile indicated in step 1 that it had no more messages to send, then the SGSN can indicate in step 12 that it should move into PS-IDLE after the expiry of a timer. The access stratum signalling connections are torn down in both cases, and the mobile's eventual states are CS-IDLE, PS-IDLE and RRC Idle.

## 4.3 Security procedures

As with any wireless communication system, there are many threats to the security of UMTS. For example, an intruder may eavesdrop on communications traffic, masquerade as an authorised user to gain free access to the network's services, or even masquerade as a network to gain information from the users. To deal with these threats, UMTS has a number of security features built into it. The security features are based on those of GSM but protect against more potential threats, and in some cases include a greater degree of protection.

The most important security features in UMTS are the ones related to network access, which protect against attacks on the air interface. There are four main aspects of network access security: user identity confidentiality, authentication, ciphering and integrity protection. We will discuss these security features in the sections that follow. The system has various other security features, which carry out tasks such as protecting the information exchanged between different networks when a mobile is roaming. These other features are beyond the scope of this book.

### 4.3.1 User identity confidentiality

It is important to keep a user's identity confidential, because an intruder can use it to help clone a new UICC and gain unauthorised access to services. UMTS does this by broadcasting a user's IMSI and IMEI as rarely as possible. Instead, it uses the TMSI and P-TMSI that we introduced in Section 4.1.

The TMSI and P-TMSI are stored in the UICC when a mobile switches off, so the mobile can use them to identify itself when it switches on again. The IMSI is only broadcast on rare occasions, such as the very first time a mobile is used. Even if the temporary identities are compromised, they are updated at regular intervals, so an intruder gains little benefit from knowing them.

### 4.3.2 Authentication

The mobile and the home network use the authentication process to confirm each other's identities, so that the network can satisfy itself that the mobile is valid, and vice-versa. The process relies on shared knowledge of a user-specific secret key, denoted K, which is stored in the home network's authentication centre (AuC) and is physically distributed to the user inside the UICC. We will describe it from the viewpoint of the packet switched domain, but it can be run from the circuit switched domain as well.

The authentication process is shown in Figure 4.8. It begins when the core network receives a suitable trigger, such as the attach request from Figure 4.7. The SGSN sends a message to the AuC (1), which identifies

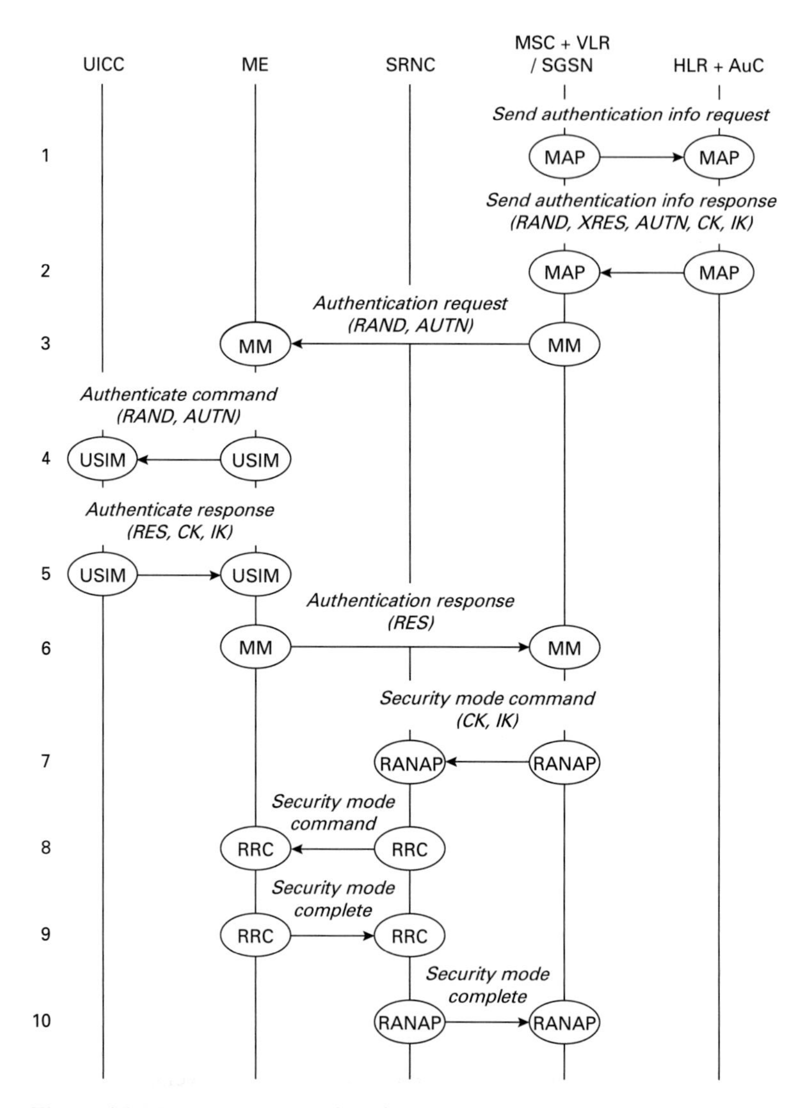


**Figure 4.8** Message sequence chart for the processes of authentication and key establishment.

the user and asks for a set of data known as an *authentication vector*. On receiving this message, the AuC thinks of a random number, RAND: this is the first part of the authentication vector, and serves as an authentication challenge that the network will send to the mobile. The AuC then looks up the secret key, and combines it with the random number to calculate the four other parts of the vector:

XRES This is the expected response to the network's challenge. It can only be computed by a mobile that knows the secret key K.

AUTN This is a guarantee to the mobile that the network is valid. It can only be computed by a network that knows two things: the secret key, and a sequence number SQN that is incremented from one authentication request to the next.

CK This is a key that will subsequently be used for ciphering, described below.

IK This is a key that will subsequently be used for integrity protection.

The calculation is done using an algorithm that is very hard to reverse: if intruders intercept the authentication vector, they cannot work out the secret key or the sequence number within a realistic time. The sequence number prevents a spoof network from recording an authentication request and replaying it, as the spoof network is unable to work out the authentication token that would come from the next value of SQN.

Once the authentication vector has been calculated, the AuC returns it to the SGSN (2). (Optionally, the SGSN can ask for several authentication vectors at a time, to reduce the number of messages that are sent to the AuC.) The SGSN sends the random number and authentication token to the ME (3), and the ME forwards them to the UICC (4).

The UICC examines the authentication token: if it indicates that the AuC knows the secret key and the sequence number, then the UICC concludes that the network is valid. The UICC then computes its response to the AuC's challenge, RES, together with its own versions of the ciphering and integrity protection keys. It returns these three values to the ME (5), which stores the two keys and returns RES to the SGSN (6). The SGSN compares the mobile's response with the expected one: if they match, it concludes that the mobile is valid.

There are just a few more tasks to do. The SGSN tells the SRNC to start ciphering and integrity protection, and tells it the keys it should use (7). In turn, the SRNC tells the ME to start those two processes (8), using the keys that the ME has previously stored. The ME and SRNC respond (9, 10), and the authentication procedure is complete.

It is worth pointing out that the authentication algorithm lies strictly within the UICC and the AuC. Inside the mobile, for example, the secret key is securely stored in the UICC and never reaches the ME; in the network, it is securely stored in the home network's AuC and never reaches the visited network. These measures ensure that the key really is a secret one. It is also worth noting that the UICC and the AuC are both under the control of the home network operator, so the network operator can use any authentication algorithm it wishes. As a result, the 3GPP specifications just include the requirements that an authentication algorithm must satisfy and an example of one such algorithm, but the example is not mandatory.

### 4.3.3 Ciphering and integrity protection

By the end of the authentication process, the ciphering and integrity protection keys have been stored in the ME and SRNC, and the two devices are ready to begin those processes.

The ciphering procedure is shown in Figure 4.9. The transmitting device uses the ciphering key CK to compute a pseudo-random stream of

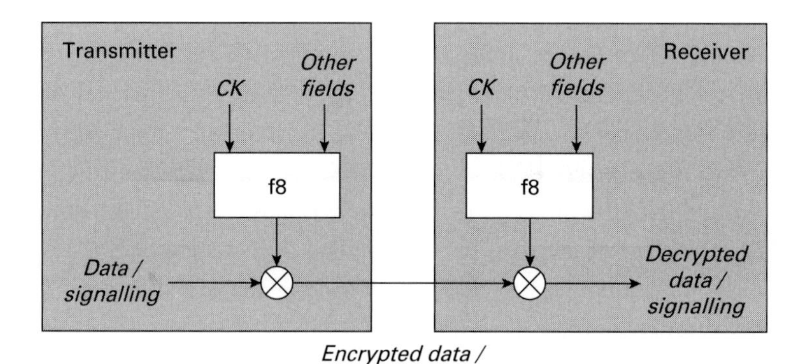


**Figure 4.9** Operation of the ciphering algorithm between the mobile and the radio access network. (Adapted from 3GPP TS 33.102.)

bits, which it combines with the transmitted data using an exclusive-or operation. The receiver carries out the same process, so as to reverse the effect of ciphering. Ciphering is recommended for the network but not mandatory: if used, it is implemented in either the radio link control or the medium access control protocol, depending on the chosen RLC mode, and applied to both data and signalling.

Integrity protection (Figure 4.10) is implemented in the RRC protocol, and is mandatory for most of the signalling messages exchanged between the mobile and the network. Using the signalling message and the integrity protection key IK, the transmitter computes a field known as MAC-I, which it appends to the message. The receiver compares the two, and finds out if they are consistent: if not, the receiver concludes that an intruder has tampered with the message, and discards it. The integrity protection algorithm also contains a replay protection procedure based on a sequence number.

Ciphering and integrity protection are implemented in the ME and the SRNC, neither of which is controlled by the home network operator. The specifications therefore include mandatory algorithms for these two processes, denoted f8 and f9. The algorithms are publicly defined, on the basis that academic researchers will find and publish any flaws before intruders can exploit them. The UMTS ciphering algorithm is intended to be exportable worldwide, under the terms of the Wassenaar agreement.

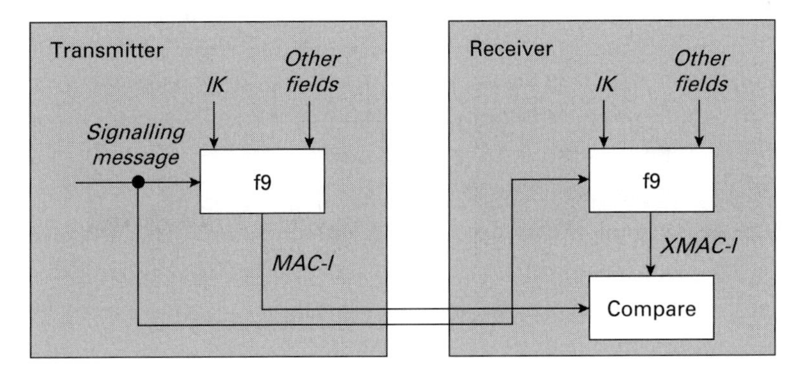


**Figure 4.10** Operation of the integrity protection algorithm between the mobile and the radio access network. (Adapted from 3GPP TS 33.102.)

Integrity protection was not implemented in GSM, while the UMTS ciphering algorithm is much more powerful than its GSM predecessor. For example, researchers have found a technique for breaking GSM's original ciphering algorithm by analysing as little as two seconds of a phone conversation. At the time of writing, no flaws in the UMTS algorithms are known that can be exploited in a realistic amount of computing time.

## 4.4 Procedures in idle mode and common channel states

We now turn to the procedures that a mobile runs after it has switched on. It turns out that the choice of procedures depends on the RRC state that a mobile is in, so we break down the description into two sections. In this section, we describe the mobile's behaviour in CELL_FACH, CELL_PCH and URA_PCH states, and in RRC Idle mode. Its behaviour in CELL_DCH state is rather different, so we will delay the discussion of that state until Section 4.5.

In the states considered here, the mobile has to carry out four main tasks. The mobile has to read the system information messages that we introduced in Section 4.2.3, to find out how the network is configured, and how to communicate with it. The mobile also has to read any paging messages that the network sends to it, so that it can receive incoming calls. As the mobile moves from one location to another, it has to measure the signal levels that it receives from the cells around it, camp on a new cell if required, and keep the network informed about its current location. Finally, in CELL_FACH state, the mobile has to transmit and receive information to and from the network. Collectively, these tasks are often known as *idle mode procedures*. This is a poor name, however, as the mobile is not necessarily idle: it can be transmitting and receiving on a common physical channel as well.

It is worth noting a few aspects of the mobile's behaviour in these states, for comparison with CELL_DCH state later on. First, the network usually controls the mobile by means of system information messages, which are the same for all the mobiles in the cell. Second, the mobile only makes a few measurements in these states, and uses most of them internally without reporting the results to the network. Third, the mobile makes its own decision about which cell it will camp on, and keeps the network informed if required. Fourth, soft handover is not supported in these states; high speed packet access is not normally supported, although CELL_FACH state can use a modified version of HSDPA from release 7 (Section 3.2.7).

### 4.4.1 System configuration

In the states we are considering, there are various possible communication paths between the mobile and the network. These are shown in the four parts of Figure 4.11.

Figure 4.11a shows the configuration in RRC Idle mode. Here, the mobile is camping on a cell, but it does not have an RRC connection or a serving RNC. From the core network's point of view, the mobile is in CS-IDLE state, and in either PS-IDLE or PS-DETACHED.

Now assume that the mobile sets up an RRC connection, and that the network places it in CELL_FACH state. (It might do this for a low data rate packet switched service like an online poker game, for example.) The

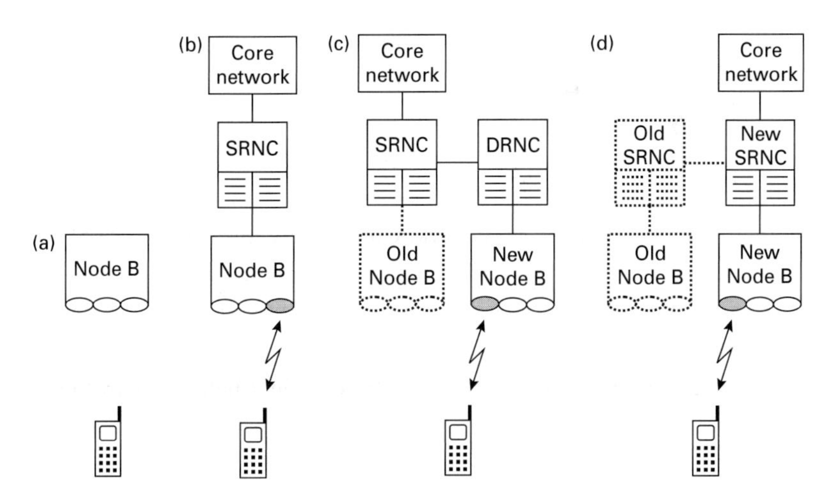


**Figure 4.11** System configurations in RRC Idle mode and common channel states. (a) In RRC Idle mode. (b) After an RRC connection setup. (c) After a cell update, with a change to a new RNC. (d) After an SRNC relocation.

system configuration is in Figure 4.11b: the mobile is still camping on a cell, but it has now been allocated a serving RNC that is acting as its sole point of contact with the core network.

Now assume that the user moves from one cell to another. Using the cell reselection procedure described below, the mobile camps on first one cell and then another, keeping the network informed as it goes. Eventually (Figure 4.11c), it may camp on a cell controlled by a different RNC. If this happens, then the new RNC takes on the role of a drift RNC, and forwards data and signalling messages over the Iur interface between the Node B and the serving RNC. Usually, the original serving RNC is retained.

If the user travels far enough, then the path from the Node B to the SRNC can get very long. If this happens, the SRNC may choose to initiate a process called *SRNC relocation* (Figure 4.11d). In this process, the SRNC functionality is transferred from the original SRNC to a new one, so as to shorten the communication path in the radio access network. This process is unusual and requires a lot of signalling, so we will not consider it further.

### 4.4.2 Reception of system information broadcasts

When a mobile is in RRC Idle mode or in one of the common channel states, one of its tasks is to read the system information messages that we described in Section 4.2.3 and act accordingly. By reading SIB 5 or 6, for example, the mobile discovers the channelisation code that the network is using for the SCCPCH. In turn, this allows the mobile to receive the paging messages described below.

To minimise its power consumption, the mobile does not have to read the system information continuously: instead, it only does so when various conditions are met. For example, it reads the system information whenever it moves into a new cell. The network can also inform the mobile about changes to the system information using a paging message; when the mobile receives such a message, it inspects a *value tag* in the master information block to find which SIBs have changed, and then re-reads the corresponding SIBs. Finally, the mobile reads some of the more rapidly changing SIBs (such as SIB 7) at regular intervals, determined by the expiry of a timer.

### 4.4.3 Paging

In CELL_PCH and URA_PCH states and RRC Idle mode, the mobile enters a mode of operation known as *sleep mode*, in which its power consumption is very low. In sleep mode, the radio access network can still send information to the mobile by means of paging messages, using the information flows that we showed in Figure 2.17d. The actual messages are sent on the *paging channel* (PCH), which is mapped onto the secondary common control physical channel (SCCPCH). The *paging indicator channel* (PICH) supports the paging process, and is only visible within the physical layer.

Figure 4.12 shows how the channels interact. When the radio access network wishes to contact a mobile, it first alerts the mobile by sending it a *paging indicator* on the PICH. The paging indicator is sent in a specific frame (a *paging occasion*), and at a specific location within that frame (a *PICH monitoring occasion*). The network then sends the mobile an RRC *paging message* on the PCH, and labels the message using the target mobile's RNTI, TMSI or P-TMSI. The paging message

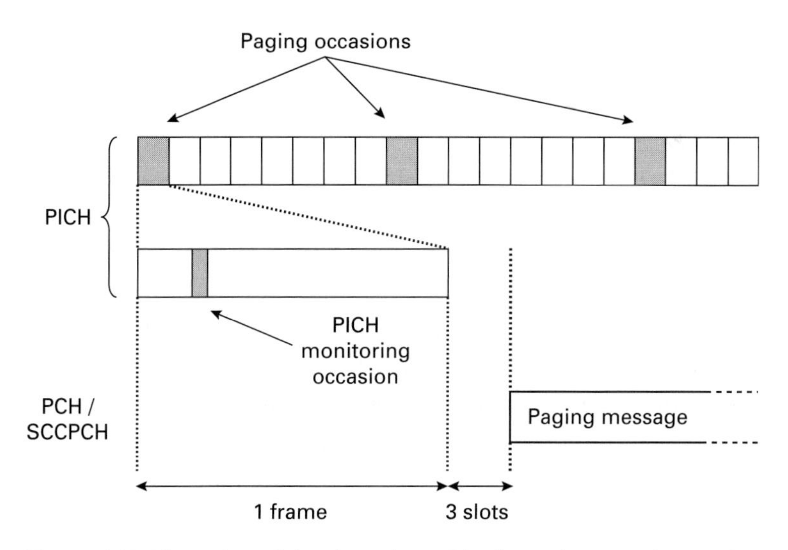


**Figure 4.12** Illustration of the channels used in the paging procedure, and the relationships between them.

lags three slots behind the end of the PICH frame, which gives the mobile time to react to the paging indicator.

If the mobile is in CELL_PCH state, then the network just sends the PICH and PCH through one cell. In RRC Idle mode or URA_PCH state, the network does not know which cell the mobile is in, so it sends the PICH and PCH through all the cells in the location area, routing area or UTRAN registration area. In CELL_FACH state, the network knows that the mobile is awake, so it doesn't bother with the PICH or PCH: instead, it just sends a paging message on the FACH.

The paging occasions are at intervals of $2^k$ frames, where $k$ is the *discontinuous reception* (DRX) *cycle length coefficient*. For paging messages originating in the core network (e.g. incoming voice calls), $k$ can lie between 6 and 9, which gives a cycle length of between 0.64 and 5.12 s. For paging messages originating in the radio access network (e.g. messages to mobiles in CELL_PCH or URA_PCH state), $k$ can lie between 3 and 9. High values of $k$ increase the delay when we want to contact a mobile, but they reduce the mobile's power consumption.

Within the discontinuous reception cycle, the network finds the exact times of the paging occasions and the PICH monitoring occasions using calculations that are based on the target mobile's IMSI. Mobiles with different IMSIs can share the same paging occasions and PICH monitoring occasions: they are said to belong to the same *paging group*.

Now consider what happens inside the mobile. In a typical implementation, a mobile might do its 3G processing using two components: an application specific integrated circuit (ASIC) for the physical layer, and a digital signal processing (DSP) chip for the higher layers. When the mobile is in sleep mode, both these chips spend most of their time in a low power state.

The mobile wakes up its ASIC during the frames containing paging occasions, and reads the PICH at the PICH monitoring occasions. (Note that the PICH is a very simple channel that is only visible to the physical layer, so it can be read using the ASIC alone.) If the mobile receives a paging indicator, then it knows that the network is about to send a paging message to a mobile in its paging group. It therefore wakes up its DSP chip, and starts to process the PCH. If the mobile's own identity matches the one in the paging message, then it passes the message up its RRC protocol and responds accordingly. Otherwise, it goes back to sleep.

The effect is that the ASIC only wakes up for the paging occasions, and the DSP chip only wakes up after a paging indicator arrives. This greatly reduces the mobile's power consumption.

### 4.4.4 Cell and network reselection

Once a mobile has camped on a cell, it runs a procedure known as *cell reselection*. Using this procedure, it measures the received signal from the current cell and the nearest neighbours, and moves to a better cell if required. The procedure is based on the cell selection procedure that we introduced in Section 4.2.4, but with a few extra features.

In RRC Idle, CELL_PCH and URA_PCH, the mobile measures the received signal-to-interference ratio (SIR) of the current cell during the frames containing paging occasions (i.e. the frames when the ASIC is already awake). If the SIR is high enough, then the mobile can continue to camp on that cell, without measuring any others. It only starts to measure neighbouring cells when one of the following criteria is met:

1. It measures cells on the same carrier frequency if:

$$Q_{\text{qualmeas}} - Q_{\text{qualmin}} \leq S_{\text{intrasearch}}$$

2. It measures cells on other carrier frequencies if:

$$Q_{\text{qualmeas}} - Q_{\text{qualmin}} \leq S_{\text{intersearch}}$$

3. It measures cells on another radio access technology $m$ (e.g. GSM) if:

$$Q_{\text{qualmeas}} - Q_{\text{qualmin}} \leq S_{\text{searchRAT}m}$$

4. It measures all neighbouring cells if, for $N_{\text{serv}}$ DRX cycles:

$$\mathrm{Q}_{\text{qualmeas}} - Q_{\text{qualmin}} \leq 0$$

The new parameters are broadcast in SIBs 3 and 4, so by reading these system information blocks, the mobile can find out when to start monitoring its neighbours. In CELL_FACH state, the ASIC is awake all the time, so the mobile measures both the current cell and its nearest neighbours continuously.

So how does the mobile find out which cells to measure? The answer lies in SIBs 11 and 12, in which the network specifies a *neighbour list* of nearby cells. This list can contain up to 32 cells on the same carrier frequency, 32 cells on other frequencies, and 32 GSM cells.

Once the mobile has measured the cells in the neighbour list, it discards any cells that are deemed unsuitable, using the criteria from Section 4.2.4. It then calculates a quality measure for each of the remaining cells: the exact measure is defined in SIBs 11 and 12, but is usually the received SIR. The mobile then moves to the best quality cell, so long as it has been better than the current cell for a long enough time.

The network can adjust the quality measure by specifying hysteresis and offset values, which encourage the mobile to stay in its current cell, and which either encourage or discourage the selection of individual neighbours. It can also apply a concept called *hierarchical cell structures* (HCS). When these are used, the network labels its cells using a priority level based on the cell size: high for picocells and low for macrocells. The mobile estimates its speed based on the number of cell reselections that it has been making. The cell reselection process is then adjusted so that slow-moving mobiles prefer high priority cells (which collectively have a high capacity), while fast-moving mobiles prefer low priority cells (to reduce the amount of signalling).

So far we have just considered cell reselection, but there is a process of network reselection as well. If the mobile is outside its home network, then in automatic mode, it searches at regular intervals for a higher priority network and, if successful, selects a suitable cell there. Similarly, if the mobile moves outside the network's coverage area, then it either selects a new network automatically, or presents a list of possible networks to the user.

### 4.4.5 Location updates

If the mobile selects a new cell, then it may need to inform the network. Its exact action depends on its current RRC state. In CELL_FACH or

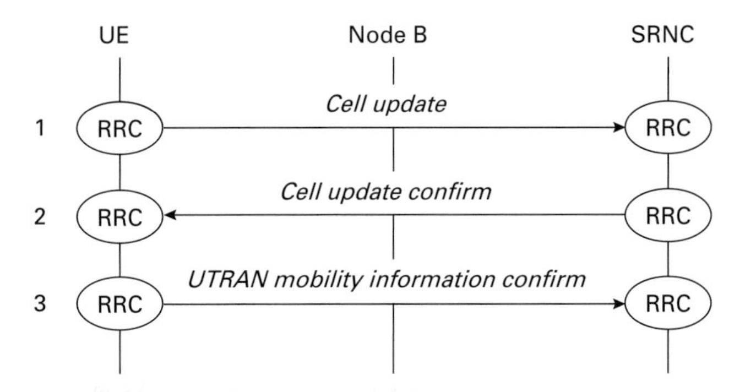


**Figure 4.13** Message sequence chart for the cell update procedure. (Adapted from 3GPP TR 25.931.)

CELL_PCH state, it tells the radio access network every time it moves to a new cell, using an RRC message called *cell update*. The messages for this procedure (Figure 4.13) are relatively simple; the only point to note is that the mobile is likely to receive a new cell RNTI as part of step 2, in which case it confirms its new identity in step 3. In URA_PCH state, the mobile reads the URA identity from SIB 2 and only tells the radio access network if it has moved to a new URA, using an RRC message called *URA update*.

In RRC Idle mode, the mobile's behaviour depends on its core network state. If it is attached to the PS domain, it reads the routing area identity from SIB 1 and, if necessary, informs the PS domain using a GMM message called *routing area update*. Otherwise, it reads the location area identity and, if necessary, informs the CS domain using an MM message called *location area update*. These messages have a similar effect to the attach request that we described in Figure 4.7, and their message sequence charts are very similar.

### 4.4.6 Radio transmission and reception in CELL_FACH state

In CELL_FACH state, unlike the others, the mobile needs to transmit and receive. Uplink transmissions are carried out on the random access channel (RACH), using the mechanism that we described in Section 3.1.2. On the downlink, the mobile monitors the FACH for occurrences of its RNTI in the MAC header. When its RNTI appears, it passes the received information to higher layers, and processes the information in the usual way.

There is a problem for mobiles that are measuring cells on other carrier frequencies: most mobiles just work on a single frequency, so are unable to receive data on one frequency and simultaneously measure a different cell on another. To get round this problem, the network uses frames known as *FACH measurement occasions*. These are frames that are computed using the mobile's IMSI, in which the network promises not to send anything to the mobile. During these frames, the mobile can measure cells on another frequency, confident that it will not miss anything on the downlink.

## 4.5 Procedures in CELL_DCH state

CELL_DCH state is used for voice calls and other types of communication over the circuit switched domain, and for high data rate packet transfers. The mobile behaves very differently in CELL_DCH state from the other states we saw earlier. First, the network takes greater control over the mobile than before, by sending it signalling messages that are specific to each individual mobile. Second, the mobile makes more measurements in CELL_DCH state than the others, and reports the results of those measurements to the network. Third, the network decides which cell, or cells, the mobile should communicate with, and tells the mobile by means of signalling commands. Fourth, soft handover and high speed packet access are both supported in this state.

In the following sections, we describe the system configuration in CELL_DCH state, and the measurements that a mobile carries out. We also review some of the procedures that can take place as a result of signalling commands from the network: active set updates, compressed mode and hard handover.

### 4.5.1 System configuration

The first important feature of CELL_DCH state is *soft handover*, which we have already seen in Section 2.2.3. In this state, the mobile can communicate with more than one cell at a time, so long as those cells are on the same carrier frequency.

The words 'soft handover' actually encompass two slightly different states. In one state, the mobile is communicating with two cells that are controlled by different Node Bs. On the downlink, the mobile receives one or more rays from each cell, and adds them together using its rake receiver (Section 3.1.5). On the uplink, the two Node Bs do their physical layer processing independently, and pass the transport blocks to the SRNC along with indications of whether those blocks have passed or failed the cyclic redundancy check (Section 3.1.6). If both have passed or failed then the result is unchanged, but if one has passed and the other has failed, then the SRNC can select the one that has passed.

In the other state, sometimes known as *softer handover*, the mobile is communicating with two cells that are controlled by the same Node B. In softer handover, the rake receiver can combine the incoming rays on the uplink as well as on the downlink, because the rays are arriving at the same physical location.

More extensive combinations are possible: for example, the words *soft softer handover* refer to a situation where the mobile is using two cells from the same Node B and a third cell from a different Node B. The *active set* is the set of cells that a mobile is currently using in soft handover: the specifications require a mobile to support active sets that contain six cells.

As shown in Figure 2.4b, the cells in the active set can be controlled by different RNCs. In this situation, the original SRNC usually stays as the mobile's point of contact with the core network, and enlists one or more others as drift RNCs. As in the other RRC states, the SRNC functionality is only moved in the event of an SRNC relocation.

It is worth pointing out that the state of soft handover is not a transient one; instead, it depends on the relative strengths of the signals that the mobile is receiving, and hence on its physical location within the network. In a typical network, roughly one third of the mobiles might be in soft handover at any one time.

Soft handover brings a number of benefits. On the uplink, it improves the cell's coverage, because it increases the signal power that the network receives from a distant mobile. It also improves the capacity, because it allows mobiles that are closer to the base station to transmit weaker signals than before, which reduces the interference that the base station receives from them. On the downlink, it improves the coverage as before. However, it has little effect on capacity: it reduces the power that each base station antenna transmits to a particular mobile, but it increases the number of transmitting antennas, so the total amount of interference stays roughly the same.

### 4.5.2 Measurements

A mobile uses a much higher data rate in CELL_DCH than in the other RRC states, so it has a much bigger effect on the network's operation. It therefore makes more extensive measurements in CELL_DCH state, and reports the results to the network so that the network can monitor its behaviour more effectively.

The first difference from the other states lies in the way that measurements are controlled. When a mobile enters CELL_DCH state, it remembers the system information that it was using in its previous state, including the measurement control information read from SIB 11 or 12. However, to minimise the mobile's processing requirements, it has no further obligation to read the system information. Instead, the network tells the mobile about any subsequent changes to the measurement control information using a mobile specific RRC message called *measurement control*.

In this message, the network can tell the mobile to make several types of measurement. As in the other RRC states, the most important are measurements of the received signal from nearby cells, which might be using the same carrier frequency, or a different carrier frequency, or a different radio access technology. However, the network can tell the mobile to measure and report several other quantities as well. One example is the downlink block error ratio, which supports the outer part of the downlink power control loop.

The other difference lies in the way the measurement results are used. In CELL_DCH state, the mobile reports the results back to the network using an RRC message called *measurement report*. Measurement reports are sent on the basis of *reporting criteria* that are defined in the

Table 4.3 *List of the measurement events that a mobile can report, with descriptions of the quantities being measured.*

| Measurement event | Description |
|---|---|
| 1A–1J | Signal strength on the current UMTS frequency |
| 2A–2F | Signal strength on a different UMTS frequency |
| 3A–3D | Signal strength on a different radio access technology |
| 4A–4B | Uplink buffer occupancy |
| 5A | Downlink block error ratio |
| 6A–6G | UE transmit power and timing offsets |
| 7A–7C | Measurements for location services |

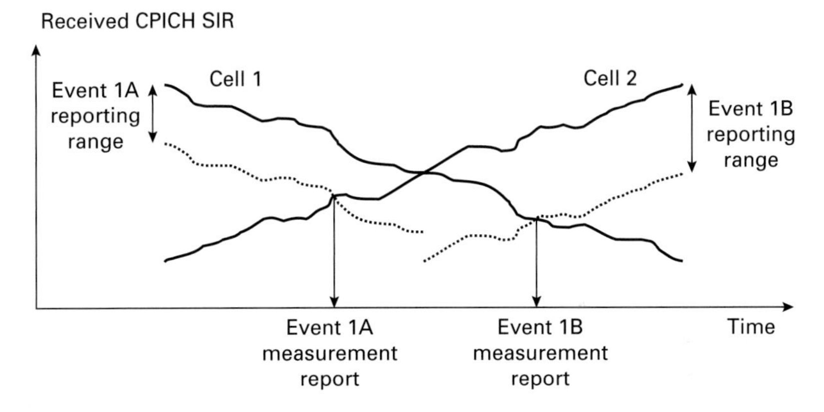


**Figure 4.14** Illustration of the measurements that are most often made in support of soft handover. (Adapted from 3GPP TR 25.922.)

measurement control message. Various different criteria can be used, such as periodic reporting, or reporting triggered by a threshold crossing event. Table 4.3 summarises the measurement quantities that are defined by the specifications, along with the associated measurement events.

A full description would take too long, so it is best to use an example. Figure 4.14 shows the measurements that are typically made in support of soft handover. At the start of the process, we assume that the active set just contains cell 1, and that the only other cell nearby is cell 2. In a measurement control message, the SRNC tells the mobile to measure the CPICH signal-to-interference ratio received from both cell 1 and cell 2. (Cell 2 is said to lie in the *monitored set*, which is the set of cells that are outside the active set but are still being measured.) It also tells the mobile to send a measurement report when either of the following events takes place:

Event 1A The signal from a monitored cell crosses upwards through a threshold that lies a few decibels below the signal from the strongest cell. (The difference is known as the *reporting range*.)

Event 1B The signal from an active cell crosses downwards through a threshold that lies a few decibels below the signal from the strongest cell.

When the signal from cell 2 crosses the first threshold, the mobile sends the SRNC a measurement report which states that event 1A has happened, and which identifies the cell responsible. The SRNC will usually react by adding cell 2 to the active set, in the manner described below. Later, the signal from cell 1 crosses the second threshold: the mobile sends a measurement report describing event 1B, and the SRNC removes cell 1 from the active set.

Various refinements are possible. To prevent the mobile from bouncing in and out of soft handover, the SRNC usually makes the reporting range for event 1B greater than for event 1A. The SRNC can also specify a time delay, so that the mobile only sends a measurement report if the signal has crossed the threshold for a certain minimum amount of time. Other measurement events can be used in support of soft handover: for example, event 1D signals a change in the best cell, and is used to trigger a change in the HSDPA/HSUPA serving cell. However, events 1A and 1B are the most important ones.

### 4.5.3 Active set updates

Soft handover is controlled using an RRC message called *active set update* (Figure 4.15), which takes a mobile in and out of soft handover

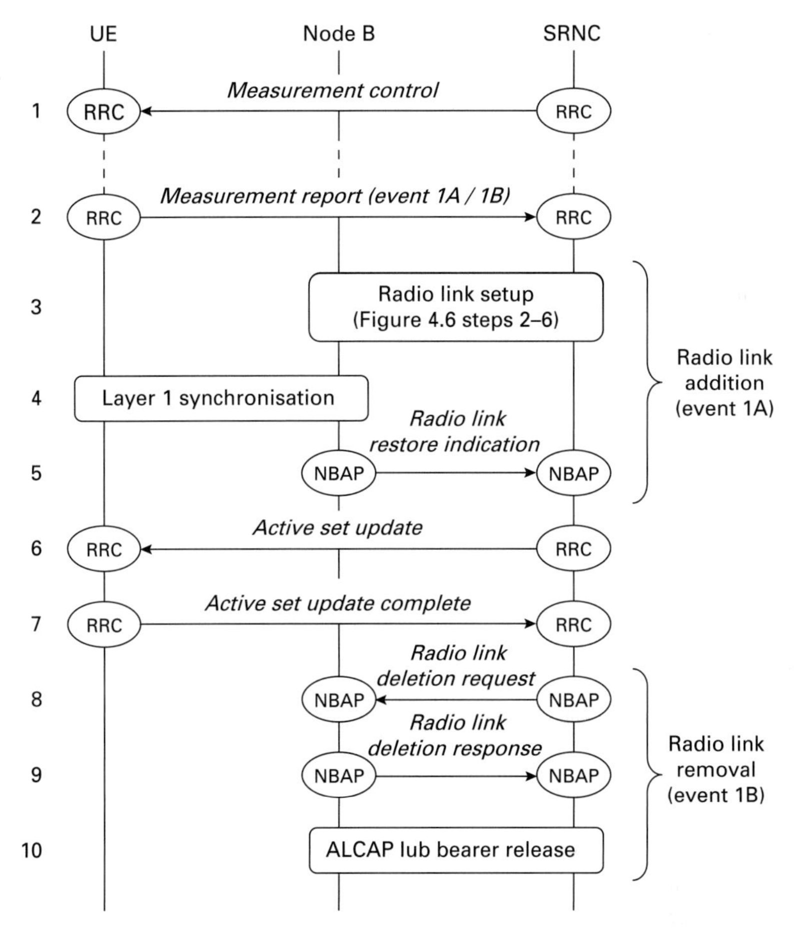


**Figure 4.15** Message sequence chart for the active set update procedure. (Adapted from 3GPP TR 25.931.)

by adding cells to and subtracting cells from the active set. The process begins when the SRNC sends the mobile a *measurement control* message (1), to tell the mobile which cells to measure and which reporting criteria to use. When one of the reporting criteria is eventually satisfied, the mobile sends the SRNC a *measurement report* (2).

If the measurement report refers to event 1A, then the SRNC usually decides to add the new cell to the active set. It starts by setting up the radio link (3), with the same messages that we saw earlier in the RRC connection setup. The Node B establishes physical layer communications with the mobile (4), and informs the SRNC (5). By the end of this process, the Node B is transmitting and receiving on the requested codes, and a transport bearer has been established between the new Node B and the SRNC.

The SRNC can now send its *active set update* message to the mobile (6). In this message, it indicates the channelisation and scrambling codes of any new cells to add to the active set, together with the scrambling codes of any cells to delete. The mobile reconfigures its physical layer, so that it starts receiving from the new cells and stops receiving from the old ones. (No change is needed to the mobile's transmissions on the uplink.) The mobile responds with an *active set update complete* (7).

If the original measurement report referred to event 1B, then the SRNC removes any old cells from the active set. This is done after communications with the mobile, to ensure that the process is free from discontinuities. In step 8, the SRNC sends a *radio link deletion request* to any Node Bs whose cells are being removed. The cell stops communicating with the mobile, and the Node B responds (9). Finally, the SRNC exchanges ALCAP messages with any outgoing Node Bs (10), to tear down the temporary virtual circuits that were being used. In practice an active set update can involve either cell addition, or cell deletion, or both at the same time.

### 4.5.4 Compressed mode

As in CELL_FACH state, the mobile may need to measure the received signal strength from cells on another frequency: either UMTS cells, or cells using another radio access technology such as GSM. In CELL_DCH state, unfortunately, the mobile is continuously transmitting and receiving. This causes a problem.

The solution to the problem is the CELL_DCH analogue of FACH measurement occasions, known as *compressed mode* (Figure 4.16). In compressed mode, the network introduces *transmission gaps* a few slots long into the downlink and/or the uplink. These can either lie in the middle of one frame, as shown, or straddle the boundary between two

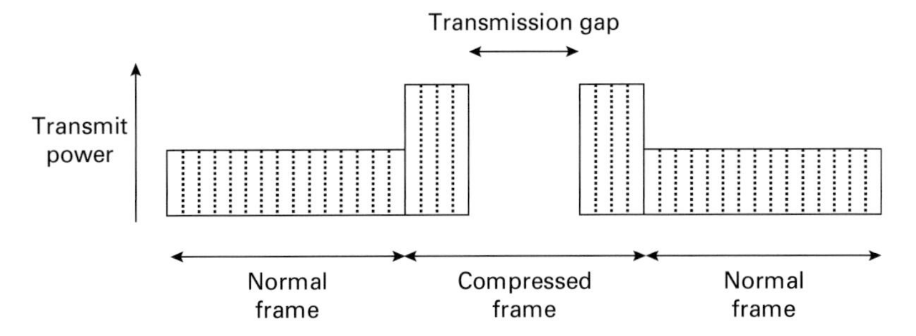


**Figure 4.16** Generation of a transmission gap in compressed mode, for the case of a real time data stream and a gap in the centre of one frame. (Adapted from 3GPP TS 25.212.)

frames. During the transmission gaps, the mobile can move to the other frequency and measure the signals received from the cells that are there. Note that the downlink transmission gaps are set up for individual mobiles: if one mobile enters compressed mode, then the base station's transmissions to the others are unaffected.

As part of its capability statements, the mobile tells the network whether it requires compressed mode on the uplink or downlink for particular types of measurement, such as measurement of a GSM cell. Downlink compressed mode is required for the same reasons as FACH measurement occasions, usually if the mobile has a single frequency receiver. The mobile can move to another frequency during a transmission gap, confident that it will not miss any data from the network. Uplink compressed mode is usually needed when measuring a GSM cell, as most mobiles are unable to transmit using UMTS and simultaneously make a measurement using GSM.

It is easy to introduce transmission gaps into a non-real time data stream, as the transmitter can just delay sending the data for a few slots. Real time streams like voice are harder. The solution is to halve the spreading factor in frames that contain a transmission gap, so as to double the data rate in the remaining slots. As shown in Figure 4.16, the transmit power is increased in the remaining slots as well, so that the same signal energy is transmitted as before. In addition, the downlink scrambling code is temporarily changed, to prevent a clash of channelisation codes with another mobile.

### 4.5.5 Hard handover

As well as soft handover, UMTS uses various types of hard handover, in which the old radio link is broken before the new one is created. The simplest example is a UMTS hard handover (Figure 4.17), which is used to move a mobile between two cells on different carrier frequencies. In the description that follows, we assume that the mobile starts in a city, and is communicating with the microcells that are installed there. If it starts to moves out of the city, then it reaches the edge of the area covered by microcells, and has to be handed over to a macrocell on another frequency. The measurement events shown in the figure are typical of the ones that a network might use, but hard handover can be triggered in other ways as well.

At the start of the sequence, the SRNC sends the mobile a measurement control message (1), which tells it to monitor the quality of its current carrier frequency. The quality measure is essentially a weighted sum of the signal-to-interference ratios from the cells in the active set. If the signal quality drops below a threshold level, then the mobile sends the SRNC a measurement report (2) describing the corresponding event, event 2D.

The SRNC now knows that the mobile is reaching the edge of the city and that a hard handover may be needed. It sends the mobile a *physical channel reconfiguration* message (3), which tells it to reconfigure its physical layer by entering compressed mode with the specified transmission gap pattern. When the mobile responds (4), the SRNC sends it a second measurement control message (5), which tells it to use the transmission gaps to measure the signal quality on the carrier frequency used by the macrocells. If the mobile reaches a state where the macrocell quality lies above one threshold and the microcell quality lies below another threshold, then the mobile sends the SRNC a second measurement report (6) describing another type of event, 2B.

This second measurement report triggers the handover. The SRNC tells the new Node B to start transmitting and receiving, in the way we have already seen (7). It then sends the mobile a second physical channel reconfiguration message (8). This tells the mobile how to reconfigure its physical layer so as to start transmitting and receiving on the new

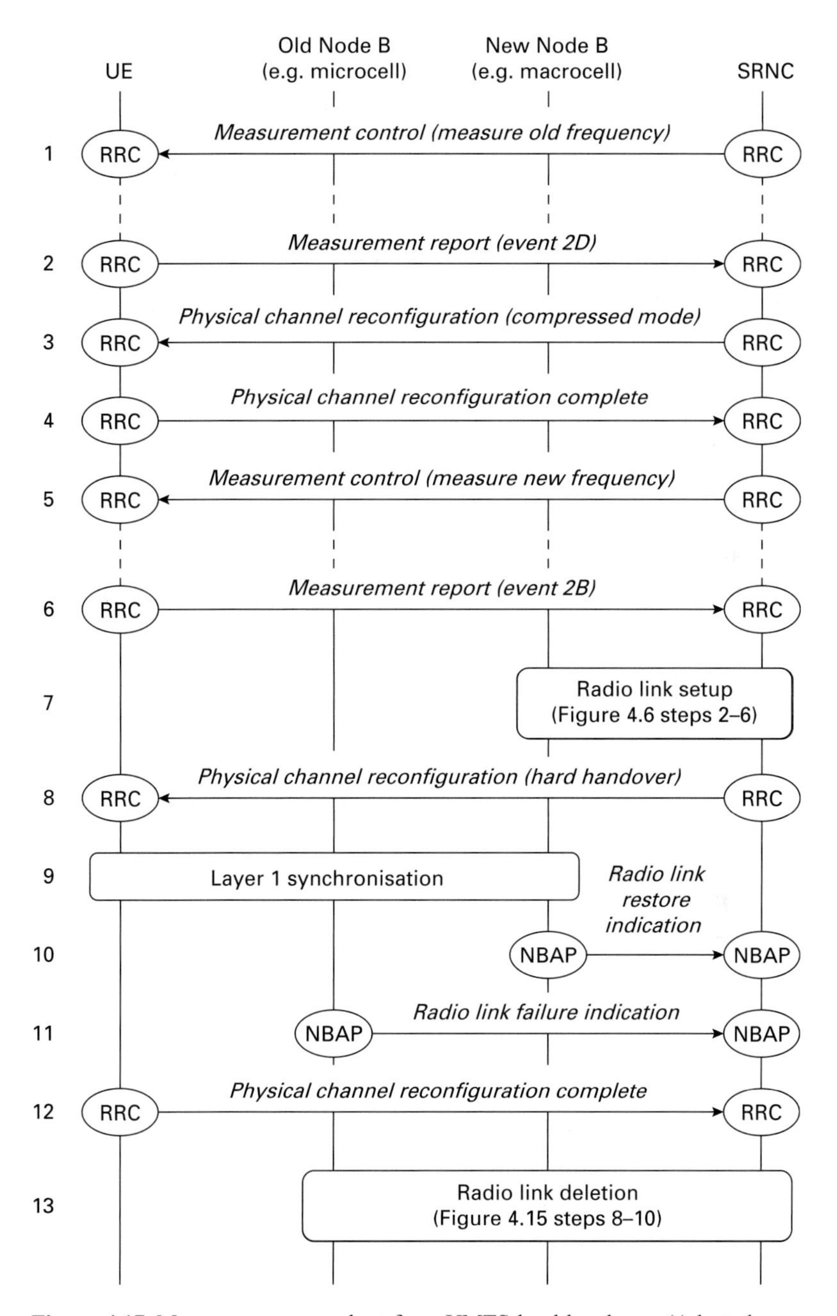


**Figure 4.17** Message sequence chart for a UMTS hard handover. (Adapted from 3GPP TR 25.931.)

frequency, so it does the job of a handover command. (If it wishes, the SRNC can put the mobile into an immediate state of soft handover on the new frequency, by specifying more than one cell.)

The mobile switches to the new frequency, and establishes physical layer communications with the new Node B (9). When this has happened, the new Node B informs the SRNC (10). Meanwhile, the old Node B sends a *radio link failure indication* (11). Once the mobile is fully set up on the new frequency, it acknowledges the handover command (12), which allows the SRNC to tear down communications across the old Node B (13).

Hard handover is less robust than soft handover, because the mobile is out of contact with the network between steps 8 and 12. If the mobile fails to establish communications on the new frequency, then it returns to the old one, and attempts to regain contact using an RRC message called *physical channel reconfiguration failure*. If the network picks this up then everything is still fine, but if the second message fails to get through then a dropped call will result.

If the mobile reaches the edge of the UMTS coverage area, then the network can hand it over to GSM. This can be done in two different ways, depending on what the mobile is doing. If the mobile is making a circuit switched call, then the process is similar to Figure 4.17, though with a few extra features. The handover begins when the SRNC sends the mobile an RRC message called *handover from UTRAN*, embedded in which is a GSM message that tells the mobile the characteristics of the destination GSM cell. The handover also requires an SRNC relocation, to move the RNC functionality from the UMTS RNC to the GSM BSC. A similar process is used for handovers from GSM to UMTS.

If the mobile is only involved in packet switched transfers then the network sends it a different message, *cell change order from UTRAN*. This looks like a handover on the UMTS side, but triggers a cell reselection once the mobile reaches GSM.

## 4.6 Power-off procedures

When you press a mobile's off button, it cannot power down right away: it has to exchange some signalling messages with the network

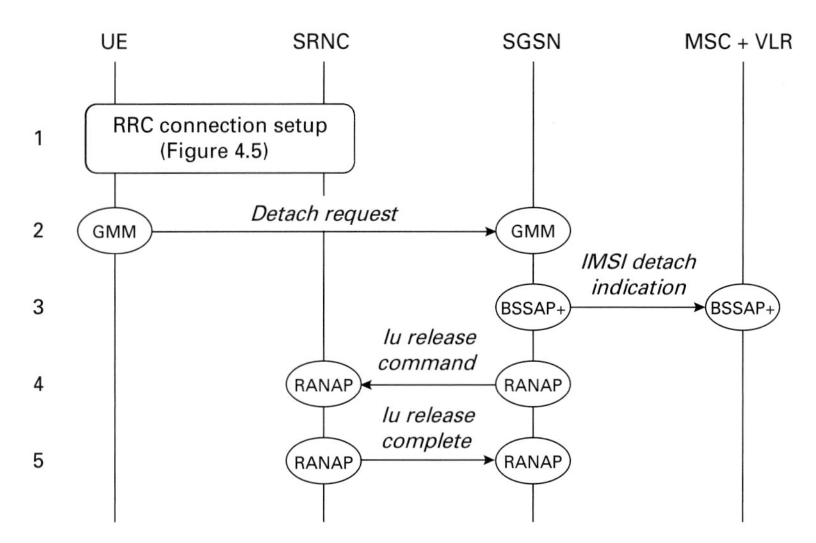


**Figure 4.18** Message sequence chart for a combined detach from the core network. (Adapted from 3GPP TS 23.060.)

first, so that it can disconnect from the network gracefully. The message sequence chart is shown in Figure 4.18, for the example of a mobile that is initially in the states CS-IDLE, PS-IDLE and RRC Idle.

Paradoxically, the mobile has to start by re-establishing communications with the radio access network (1). In its RRC connection request, it indicates that the reason is a wish to detach from the core network. The target RNC identifies itself as the mobile's SRNC, and places the mobile in (usually) CELL_FACH state.

The mobile then composes a GMM message called *detach request* (2). This indicates the type of detach to perform (in this case, a combined detach with a switch-off), and the mobile's P-TMSI. The message is sent to the SGSN using an RRC initial direct transfer and a RANAP initial UE message, to re-establish the signalling connections between the mobile and the core network. As soon as it has sent the message, however, the mobile can switch off. Meanwhile, the SGSN forwards the mobile's detach request to the MSC (3).

The SGSN now has to tear down the Iu-PS signalling connection that has just been set up. It does this by sending an *Iu release command* to the SRNC (4). The SRNC responds (5), and the SRNC and SGSN can now tear down both sides of the Iu-PS signalling connection, and the network's side of the RRC connection. (No further message is required to the mobile, as the mobile is already switched off.) The CS domain can detach the mobile as soon as it receives the detach request from the SGSN. No Iu release command is required there, as there is no Iu-CS signalling connection to tear down.

There are many possible variations. If, for example, the mobile has any active packet switched communications with the outside world, then these are torn down as part of the procedure. If the mobile's battery runs out, then it stops sending periodic location or routing area updates to the core network. When a timer expires, the network notices that this has happened, and it can detach the mobile.

# 5 Services and their implementation

Having described the internal operation of UMTS in Chapter 4, we can go on to consider how the system provides services to the user. We begin by explaining how services are classified and how the network provides the user with the quality of service required. We then give a detailed description of the two most important services that UMTS provides: voice and the general packet radio service (GPRS). We focus on the signalling messages that set up, manage and tear down voice calls and data transfers, and also on the mechanisms that are used to transfer information between the mobile and the end device.

The second half of the chapter is a shorter account of the other services provided by UMTS. This account is in two parts. The first part covers the other services that are of interest to the user, such as the short message service (SMS) and the multimedia messaging service (MMS). The second part covers the toolkits that application developers can use to build up higher level services. The chapter closes with an overview of the procedures that are used for charging and billing.

## 5.1 Service classification

Ultimately, the purpose of UMTS is to provide services that the end user will pay for. The services defined by the 3GPP specifications fall into four categories.

*User services* define both the data transport mechanism and the application software, so they provide a complete end-to-end service for the user. Voice is a user service, for example, because it defines both the protocols that transport the speech information, and the codecs that encode and decode it. (Circuit switched user services are also known as *teleservices*.) In contrast, *bearer services* just define the mechanism used for data transport, so they can be used by any overlying application. GPRS is a bearer service, because it just transports packet data between the mobile and a server, without caring whether the data are documents, web pages or emails.

Most of the services described in Sections 5.3 to 5.5 are user services, the main exceptions being GPRS and the circuit switched data service. However, the distinction between the two types of service can be rather blurred in practice. For example, SMS is normally considered as a user service that encodes, delivers and decodes text messages. However, it can also be used as a bearer service, for the delivery of other types of data such as mobile ringtones.

*Supplementary services* have no independent function: instead, they support or enhance the user services, notably voice. Examples include call barring, call forwarding and call waiting. Finally, *toolkits* are low level services that application developers can use, to build higher level applications that are of interest to the user. For example, location services allow the network to determine the physical location of the user, for use by applications such as emergency calls.

## 5.2 Quality of service

Before describing the individual services listed above, we need to explain the concept of *quality of service* (QoS). This encompasses a number of issues such as data rate, error rate and delay, which collectively determine whether or not the user is satisfied with the service that the network provides. In this section we describe the parameters that are used to quantify quality of service, the way in which the mobile and network negotiate those parameters, and the way in which the network eventually implements them.

### 5.2.1 QoS parameterisation

UMTS is designed to provide a good quality of service for the UMTS bearers that we introduced in Figure 2.16, which convey information between the mobile termination and the far end of the core network. Each UMTS bearer is associated with a number of QoS attributes, which are summarised in Tables 5.1 and 5.2.

Table 5.1 *List of UMTS traffic classes, together with their quality of service requirements and some example services.*

| Traffic class | Error rate | Delay | Delay variation | Example |
|---|---|---|---|---|
| Conversational | – | Low | Low | Phone call |
| Streaming | – | – | Low | Streaming video |
| Interactive | Low | Low | – | Web browsing |
| Background | Low | – | – | Email |

Table 5.2 *Example quality of service attributes for a voice call and a web browsing application.*

| QoS attribute | Voice call | Web browsing |
|---|---|---|
| Traffic class | Conversational | Interactive |
| Maximum bit rate | 12.2 kbps | 64 kbps |
| Guaranteed bit rate | 4.75 kbps | – |
| SDU error ratio | $7\times10^{-3}$ | $10^{-6}$ |
| Residual bit error ratio | $10^{-4}$ | $6\times10^{-8}$ |
| Transfer delay | 100 ms | – |
| Traffic handling priority | – | 2 |

The most basic QoS attribute is the *traffic class*, shown in Table 5.1. This groups the data streams into four classes, depending on their broad requirements for error rate and delay. In two-way conversational services such as phone calls, for example, the user requires an end-to-end delay that is low and constant, but can tolerate a higher error rate than in many other situations. One-way streaming downloads are similar, except that a large absolute delay is much less of a problem as the user only notices it on start-up. Web browsing and emails require a low error rate but again have different delay requirements: a web client requires a low round trip delay, while an email client has few concerns beyond reliable delivery of the information.

About 15 other attributes provide finer detail: Table 5.2 lists the most important ones and gives some example values. The maximum bit rate is the highest bit rate that the network will ever supply, and determines the spreading factor that the radio access network will use. The guaranteed bit rate is the minimum bit rate that the network promises to supply: some of the other attributes (such as delay) are not guaranteed for bit rates above this figure.

The SDU error ratio is the fraction of application layer service data units that are lost or are labelled as erroneous, typically because they have failed a cyclic redundancy check on the air interface. It is different from the residual bit error ratio, which is the bit error ratio in SDUs that have passed all the error detection stages, but which may still contain errors due to imperfections in the CRC algorithm.

The transfer delay is the 95th percentile delay between the mobile termination and the far end of the core network. It is only used for conversational and streaming services, and would be much lower for the former. In the case of interactive services it is replaced by the traffic handling priority, which is a priority level (1 to 3) that expresses the delay in a more qualitative way.

Our description has focussed on the UMTS bearer, but this does not completely reflect the user's experience as it leaves out two parts of the system: the path between the terminal equipment and the mobile termination, and the path in the external network. The former is not usually a problem, but the latter can be an important source of degradation. The biggest issue is delays in external packet switched networks: these are beyond the control of early releases of UMTS, but can be controlled by the IP multimedia subsystem that will be discussed in Chapter 6.

### 5.2.2 QoS negotiation

The next issue is one for the control plane: how do the mobile and the network agree on the QoS parameters to use, and how does the network map those parameters onto the low level data streams? The basic procedure is as follows; we will see some details when we discuss voice calls and GPRS below.

The procedure begins when the mobile sends a message to the core network, asking it to set up a data stream. As part of this message, it specifies the requested QoS, using either the same attributes that describe the UMTS bearer, or other attributes that can easily be mapped onto them. The MSC or SGSN consults its copy of the user's subscription details, and decides whether the user can receive the requested QoS. If the user has not paid for the requested service, for example, then the QoS request can be reduced or refused. It also consults admission control functions in the core and radio access networks. These functions measure the loads in the two networks and decide whether the user should be admitted.

If everything is satisfactory, then the network maps the UMTS bearer onto the lower level bearers, dividing up QoS attributes such as error rate and delay in the most appropriate way. At the end of this process, the requested attributes are implemented using parameters that describe the low level data streams. On the air interface, for example, the error rate is determined by parameters such as the number of CRC bits, and whether or not retransmissions are used.

Once everything has been set up, the network sends a reply to the mobile. This includes the QoS that the network will offer, which may be worse than the QoS requested, and which the mobile may either accept or reject. Note that the offered QoS is not an unconditional promise: if a cell gets congested later on, for example, then the radio access network may be unable to deliver the data rate that it has previously guaranteed.

### 5.2.3 QoS management

The final issue is one for the user plane: how does the network provide the user with the QoS that it has offered? In a circuit switched network we reserve a precise bit rate for every data stream, typically using time or frequency division multiplexing, which makes it easy to guarantee the allocated bit rate to each data stream. However, the situation in a packet switched network depends on the transport protocol used.

The ATM protocol stack has quality of service mechanisms built into it from the beginning, which ensure that users get the bit rates and delays they have requested. This is not the case for IP: the Internet, for example, is notoriously prone to low bit rates and long delays at times of heavy load. The solution is to add some enhancements to IP, normally using a technique called *differentiated services* or *DiffServ*. In Chapter 2, we introduced a number of protocol stacks that supported IP-based transport: these actually use IP enhanced with DiffServ, to provide the user with the quality of service that would otherwise be missing.

## 5.3 Voice calls

The oldest and most important service in a mobile telecommunication system is the voice call. In this section, we will describe the signalling messages that are used to manage voice calls, and the processes that are used to encode, transport and decode the digital signal. We will also give an overview of the supplementary services that modify how voice calls behave.

### 5.3.1 Setup of a mobile originated call

In release 99, voice calls are managed using signalling messages that are written and transported using the protocol stack shown in Figure 5.1. This combines the protocol stacks for the PSTN, Iu-CS, Iub and Uu interfaces that we saw in Figures 2.9b, 2.12a, 2.12c, 2.14a and 2.14b.

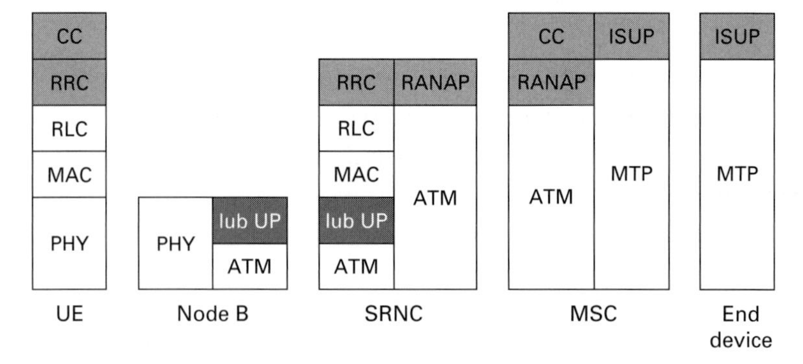


**Figure 5.1** Control plane protocol stack used to manage a voice call in release 99.

In the core network and the outside world, the signalling messages are written using the ISDN user part (ISUP), and are transported using the message transfer part (MTP) of the SS7 protocol stack. Between the MSC and the mobile, the messages are written using the UMTS call control (CC) protocol, and are transported using RRC and RANAP direct transfers. The MSC translates the messages between the ISUP and CC protocols, to ensure that the mobile can communicate successfully with the outside world.

The figure leaves out the information flows between the protocols, which are similar to the ones we saw in Chapter 2. It also assumes that the mobile is just communicating with a single Node B and a single RNC. Soft handover or the use of a drift RNC would cause changes similar to those in Figure 2.19. Later releases support other protocol options: from release 5, for example, the radio access network can transport information using IP.

To illustrate these signalling messages, Figure 5.2 shows the release 99 message sequence chart for a mobile originated call. We assume that the mobile starts in CS-IDLE state and RRC Idle mode, so it first needs to set up an RRC connection (1). In the connection request, it indicates that the cause is a mobile originated conversational call, and the serving RNC reacts by placing it in CELL_DCH state.

The mobile now has to establish a signalling connection with the MSC. To request this, it sends the MSC a message called *CM* (connection management) *service request* (2). The message includes the reason for the request, which in this case is a mobile originated call. Optionally, the MSC can run the security procedures that we saw in Chapter 4 (3). If it does so, then the mobile interprets the last security message it receives (the RRC message security mode command) as an implicit acknowledgement of its request from step (2). Otherwise, the MSC sends the mobile a separate message called *CM service accept*, which does the acknowledgement explicitly.

Now that it has a signalling connection, the mobile can send the MSC a call control message known as *setup* (4). This asks the MSC to establish the call, and includes parameters such as the called party's phone number and the requested bit rate. The MSC checks the request against the user's subscription details and the load in the core network,

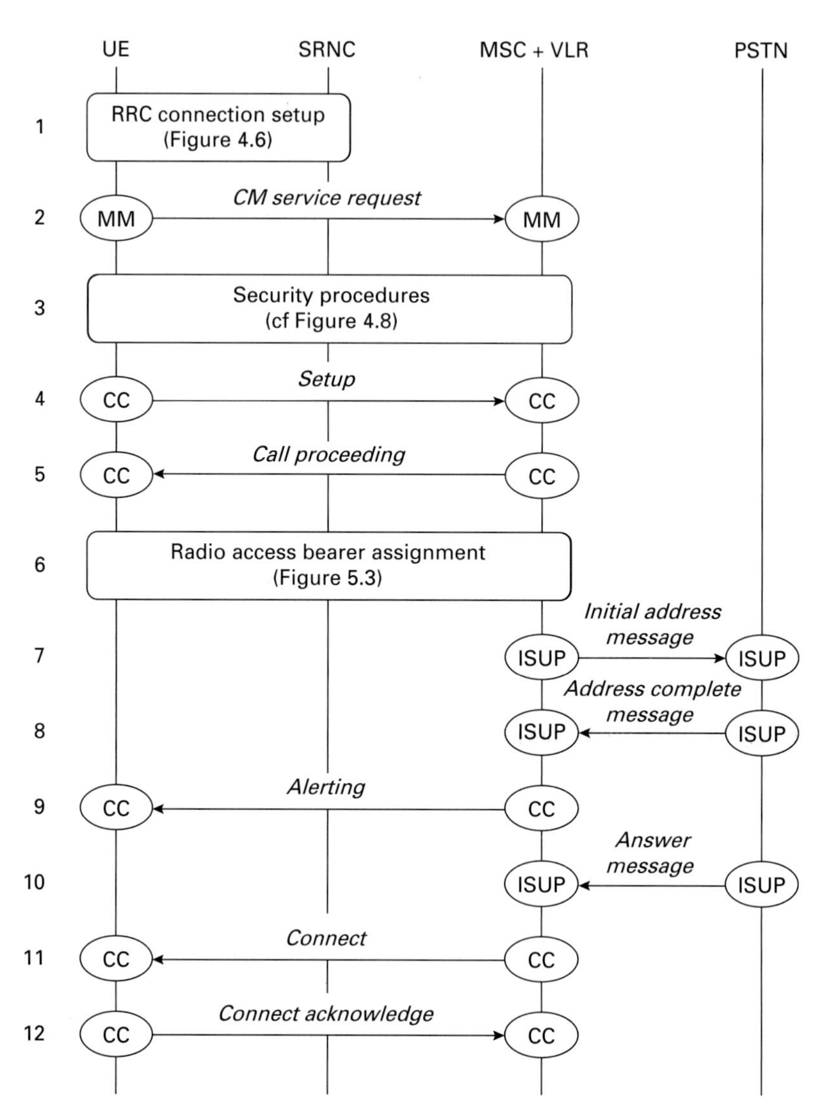


**Figure 5.2** Message sequence chart for mobile originated call setup in release 99.

as described earlier. If everything is satisfactory, it replies to indicate that the call is being processed (5).

At this point, the MSC translates the user's parameters into the QoS attributes that describe the UMTS bearer, and breaks those down into attributes for the core network bearer and the radio access bearer. It then tells the SRNC to set up the radio access bearer (6), in a self-contained process that will be described in the next section. By the end of this step, the network has set up all the bearers that will be needed to transport the call, and all the underlying data streams.

The MSC can now start to communicate with the outside world. It does this by translating the mobile's setup request into an ISUP *initial address message*, which it sends to the target device (7). The phone starts to ring, the target device replies (8), and the MSC informs the mobile that the phone is ringing (9). When the user answers the phone, the target device sends another ISUP message to the MSC (10). The MSC informs the mobile as before (11), and the mobile replies (12). The call can now start to flow.

The procedure for a mobile terminated call has a few extra steps. The originating network begins by sending an ISUP initial address message to a gateway MSC in the mobile's home network: the GMSC has to exchange a few messages with the HLR, to identify the mobile's MSC and forward the incoming message there. If the mobile is in CS-IDLE state, then the MSC only knows its current location area, so it sends a RANAP paging message to all the RNCs in the location area. The RNCs react by sending an RRC paging message through all the cells they are controlling, and the mobile replies by requesting an RRC connection. From that point on, the message sequence is very like the one for the mobile originated call, though with a few natural changes like reversing the direction of the CC setup message.

Both procedures are much the same in release 4, except for the introduction of extra messages in which the MSC server configures the media gateway.

### 5.3.2 Radio access bearer assignment

In describing voice call setup above, we skipped over the messages that set up the radio access bearer (RAB) between the mobile and the MSC. These are shown in Figure 5.3. To start the process, the MSC sends the SRNC a RANAP message called *RAB assignment request* (1), which includes the QoS attributes that the MSC has computed for the radio access bearer. The SRNC checks the request against the load in the

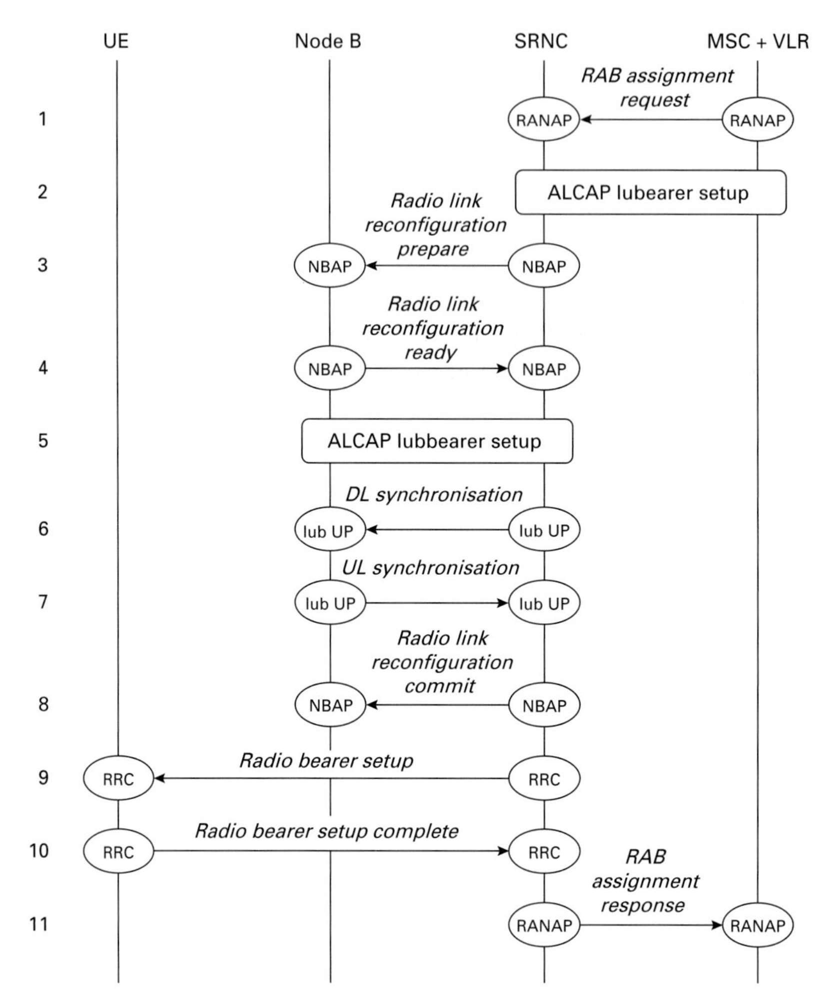


**Figure 5.3** Message sequence chart for radio access bearer assignment. (Adapted from 3GPP TS 25.931.)

radio access network, by examining the uplink noise rise and downlink transmit power in the cells that the mobile is using. If it is willing to set up the call, it translates the QoS attributes into the parameters that describe the underlying bearers and data streams. It then sets up the temporary virtual circuits that will transport the voice call over Iu (2).

The next few steps are similar to those for radio link setup, though not identical. In step 3, the SRNC tells the Node B how to modify its existing radio link, to exchange data with the mobile instead of just signalling. (For example, the message includes the new transport format combination sets for the uplink and downlink.) The Node B does not actually make the changes yet, as in this example, the reconfiguration time will be synchronised between the mobile, Node B and SRNC. Instead, the Node B simply replies to the SRNC (4). The SRNC and Node B also exchange ALCAP and synchronisation messages, in the way we have previously seen (5 to 7). Finally, the SRNC tells the Node B the connection frame number (CFN) in which to apply the changes described earlier (8). If the mobile is in soft handover, then the network repeats steps 3 to 8 for each of the Node Bs that are involved.

Once the Node B has been configured, the SRNC can send the mobile a message called *radio bearer setup* (9). This tells the mobile how to modify its radio link for the exchange of data as well as signalling, and tells it the CFN in which to apply the changes. In step (10), the mobile replies. When the requested CFN arrives, the mobile, Node B and SRNC update their radio bearers at exactly the same time. This avoids any confusion that would arise if, for example, the Node B started to transmit on the new transport format combination set, while the mobile was still receiving on the old one. Once this has been done, the SRNC acknowledges the MSC's original request (11), and the MSC can move on to the next stage of setting up the call.

### 5.3.3 Transport of voice calls

Figure 5.4 shows the release 99 protocol stack used to transport voice calls, and the corresponding data flows.

Let's look at the downlink path, from the end device to the mobile. As described in Chapter 2, the application digitises the speech signal with a sample rate of 8 kHz and a resolution of 8 bits. The digitised signal is transported to the MSC using pulse code modulation (PCM), at a rate of 64 kbps. In the MSC, the *adaptive multi rate* (AMR) codec compresses the signal, so as to reduce the load on the air interface. The AMR codec, and the resulting data streams, are the main topic of this section.

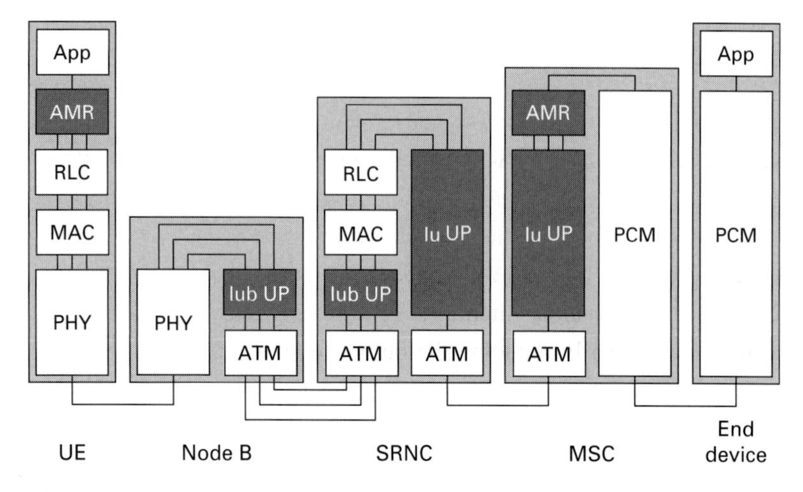


**Figure 5.4** User plane protocol stack for a voice call in release 99, together with the data flows.

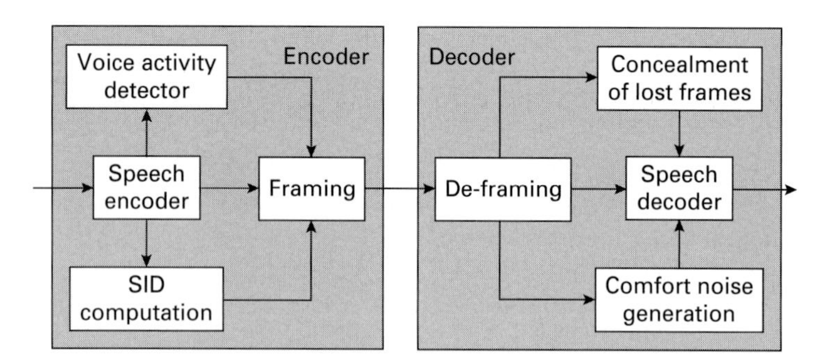


**Figure 5.5** Block diagram of the AMR codec. (Adapted from 3GPP TS 26.071.)

Figure 5.5 shows a block diagram of the AMR codec. In the encoder, the digital signal is compressed to one of eight different bit rates, from 4.75 to 12.2 kbps, depending on the AMR mode that is currently selected. (A high bit rate results in a high quality signal, but a low network capacity.) The compressed signal is then organised into 20 ms frames for transmission. At the same time, a voice activity detector determines whether or not the user is talking; if not, the compressed frames are replaced by a parameterised description of the background noise known as a *silence information descriptor* (SID). This has a bit rate of 1.95 kbps, and has two purposes: it prevents the background noise from switching on and off, and it reassures the speaker that the listener is still there.

The receiver inspects each frame, determines whether it contains speech or an SID, and decodes the speech or generates comfort noise as required. Some of the received frames may be missing or labelled as erroneous, usually because they have failed the CRC on the air interface. The receiver handles them first by repeating the preceding frames, and then by gradually reducing the output level to zero.

The output of the encoder is a parameterised description of the original speech signal. Some of the encoded bits are very important, in the sense that the reconstructed signal will be badly degraded if they are received incorrectly, while others are less important. To handle this issue, the encoder groups the compressed bits into three classes, placing the most important bits in class A, less important bits in class B, and the least important of all in class C. (At 12.2 kbps, for example, each 20 ms frame contains 81 bits in class A, 103 in class B and 60 in class C, to make the expected total of 244. SID frames contain 39 bits in class A, and none at all in classes B or C.) These three classes correspond to the three data flows in Figure 5.4.

Between the mobile and the SRNC, the different classes are mapped onto different transport channels, so that they can receive different amounts of error protection. For example, the class A bits are usually the only ones protected by a CRC. Between the mobile and the Node B, the transport channels are multiplexed onto a single physical channel, in which the frames are labelled using the transport format combination indicator (TFCI) that we introduced in Chapter 3. By varying the TFCI, we can switch between speech, SID fields and no speech signal at all, and we can also vary the selected AMR mode.

On the Iu interface, the three classes are handled using different *RAB sub-flows*, using a special mode of the Iu user plane protocol called *support mode for predefined SDU sizes*. The protocol multiplexes the bits from each sub-flow into a single Iu packet, and labels each packet with an *RAB sub-flow combination indicator* (RFCI) that works in much the same way as the TFCI.

Later specifications include a number of enhancements. From release 4, the core network can transport the speech information using ATM or IP in place of PCM. These packets can be in compressed form, so the AMR no longer needs to be implemented in the MSC; instead, the speech information can be compressed over the full end-to-end path between the mobiles, in a mode known as *tandem free operation*.

### 5.3.4 Supplementary services

*Supplementary services* (SS) enhance or modify the basic user services, by offering features such as call barring, call forwarding and call waiting. The services are managed by signalling messages that are exchanged between the mobile and the HLR. These messages provide service specific functions to the user, such as registration, deregistration and interrogation of the current service status. On the air interface, the signalling messages are handled by a separate supplementary service protocol that runs in parallel with mobility management and call control. Within the core network, the messages are handled by MAP.

Some of the services are subdivided, depending on the circumstances in which they are invoked. For example, the call forwarding service actually comprises four separate services, in which the call can be forwarded unconditionally, when there is no reply, or when the user is busy or not reachable.

As an example, the call forwarding unconditional service might be used to redirect all incoming calls to voicemail. When the user selects the service, the mobile sends an SS *register* message to the MSC, which specifies the service being registered and includes the voicemail number that has been read from the UICC. In turn, the MSC sends a MAP *register_SS* message to the HLR. When an incoming call arrives, the HLR inspects the user's supplementary service registration, and forwards the call to the voicemail number instead of the mobile's MSC.

Supplementary services are precisely defined by the UMTS specifications, so they cannot be modified and are therefore rather inflexible. There are, however, other ways in which third party developers can add extra features into the system, and we will cover those when we discuss toolkits in Section 5.6.

## 5.4 GPRS

The *general packet radio service* (GPRS) exchanges data streams, such as emails, web pages and data files, between the mobile and a server. The data are delivered over the core network's packet switched domain, using the packet switched transport techniques that we introduced in Chapter 1. This allows the network to bill its users for the amount of data transferred, not for the amount of time they have been connected.

In this section, we describe the signalling messages that manage the data transfers, and the techniques that are used to transport the data over the packet switched domain. First, however, we need to describe the system architecture used for GPRS, and introduce some new terminology.

### 5.4.1 System configuration

The purpose of GPRS is to exchange packet data between the mobile and a server. The server can be controlled by the network operator or a third party, and can be connected to a gateway GPRS support node (GGSN) either directly or across a packet data network like the Internet.

The system configuration for GPRS is shown in Figure 5.6. The server communicates with a GGSN in the PS domain, and together the two are identified using an *access point name* (APN). This is a similar concept to an Internet domain name, but has two parts. The *APN network identifier* specifies the server or data network that is connected to the GGSN. Many network operators use two different network identifiers to provide access to the operator's own services and to the Internet; others use different network identifiers for contract and prepaid users. The *APN operator identifier* specifies the PLMN that the GGSN is located in, by means of its mobile country code and mobile network code. The operator identifier is optional, in the sense that the core network can provide it, if the user has not already specified it.

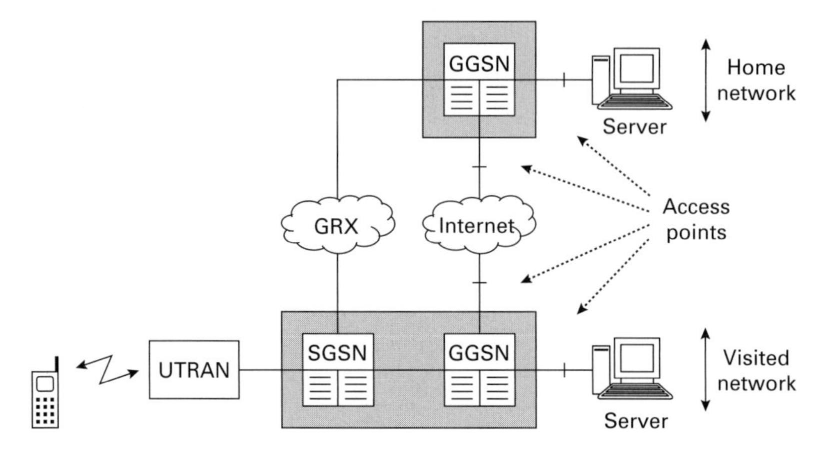


**Figure 5.6** Architecture of the general packet radio service (GPRS) for a roaming mobile.

A typical APN is internet.mnc123.mcc456.gprs. Here, the network identifier specifies access to the Internet, while the operator identifier specifies a GGSN in the network defined by mobile country code 456 and mobile network code 123. To reach a particular server on the Internet, the user would first select an access point name like this one, and then submit the required domain name.

Data can be transported between the mobile and the server using various types of *packet data protocol* (PDP). UMTS currently supports three types of packet data protocol: IP version 4, IP version 6 and the point-to-point protocol (PPP). IP version 4 is the most common one: it uses the familiar 32 bit addresses such as 192.168.0.1. Due to a shortage of IPv4 addresses, a mobile does not normally have a permanent IP address of its own; instead, the UMTS network allocates it a temporary IP address that is private to that network and invisible to the outside world. The long term solution will be to replace IP version 4 with IP version 6, which uses 128 bit addresses instead; unfortunately the transition is proving difficult and is only taking place slowly. PPP is the protocol commonly used for dial-up access to the Internet.

Within the UMTS network, there is a separate, lower level transport process that routes data packets between the SRNC, SGSN and GGSN. This process is known as *tunnelling*, and always uses IP-based transport. The tunnelling process is described by a data structure known as a *PDP context*, which is stored in the network nodes and contains the information needed to route packets between them with the correct quality of service. The PDP context parameters include the external network's access point name, the mobile's PDP address (used to route packets to and from the server), the IP addresses of the SRNC, SGSN and GGSN (used by the tunnelling process), and the QoS attributes. In the terminology used by GPRS, the process of setting up a data transfer between the mobile and the GGSN is known as *PDP context activation*, and we will describe this process in the next section.

If the mobile is roaming, then the SGSN is always located in the visited network. The GGSN is usually located in the home network, but may be in the visited one if the home network permits access to the visited network's services. The home and visited networks communicate through a multi-operator virtual private network known as the *GPRS roaming exchange* or *global roaming exchange* (GRX).

## 5.4.2 PDP context activation

Figure 5.7 shows the control plane protocol stack used for the signalling messages that control a GPRS packet data transfer.

The signalling flows are very like the ones we saw earlier for voice. Between the SGSN and the GGSN, the GPRS tunnelling protocol

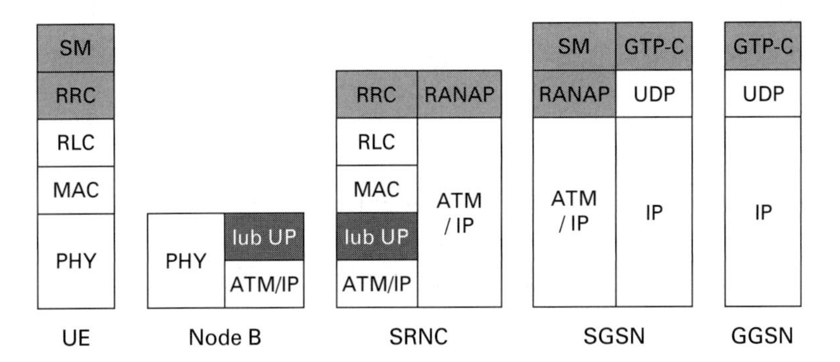


**Figure 5.7** Control plane protocol stack used for signalling flows in GPRS.

control part (GTP-C) handles signalling messages that set up, manage and tear down the data streams. These messages are transported using IP. Between the SGSN and the mobile, signalling messages are handled by the session management (SM) protocol, and are transported using RRC and RANAP direct transfers. For the moment we ignore any signalling messages that might be exchanged between the GGSN and the server: we will discuss those messages briefly below.

The message sequence chart for mobile initiated PDP context activation is shown in Figure 5.8. As in the case of the voice call, the full set of messages is rather complex, but the figure is greatly simplified

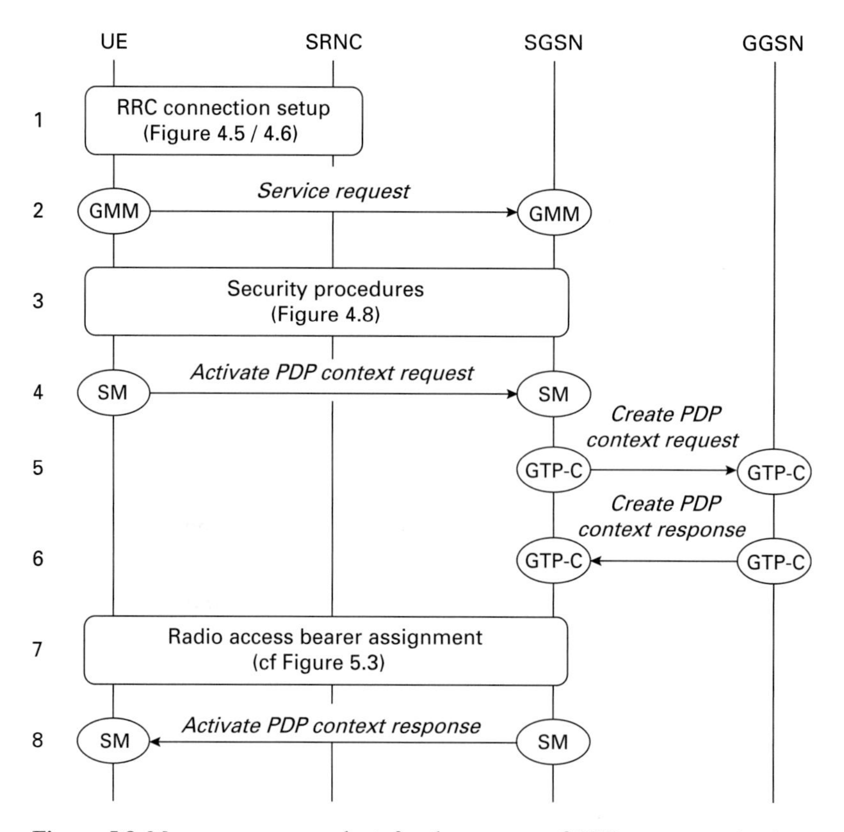


**Figure 5.8** Message sequence chart for the process of PDP context activation. (Adapted from 3GPP TS 23.060.)

because we have come across most of them before. As before, we assume that the mobile is starting in RRC Idle and PS-IDLE.

The mobile's first step is to set up an RRC connection with the radio access network (1). The reason for the request might be an originating interactive call, and the SRNC might respond by placing the mobile in CELL_DCH state. The mobile then has to establish a signalling connection with the PS domain. It does this by sending the SGSN a message called *service request* (2). As part of the message, it indicates its current P-TMSI and routing area identity, together with the reason for the request. The SGSN can initiate the security procedures (3), and it then places the mobile in PS-CONNECTED state. As in the circuit switched case, the mobile interprets the reception of an RRC security mode command message as an implicit acknowledgement of the service request.

The mobile now sends an SM message called *activate PDP context request* to the SGSN (4). As part of this message, it specifies the APN of the server it wishes to contact, the requested quality of service, and the PDP type that it wishes to use. If it is using IP and has a static IP address, then it also specifies its address here; if it is using PPP or has no IP address, then it leaves the field blank.

The SGSN now looks up the requested APN in a domain name server, which returns the IP address of the corresponding GGSN. It also checks the PDP context request against information such as the user's subscription details and the current network load. It can reject the request or reduce the quality of service if (for example) the user's subscription does not support it, or if the GGSN lies outside the user's home or visited network, or if the network load is too high. If the request is valid, the SGSN sends the GGSN a GTP-C message called *create PDP context request* (5). This includes the information in the mobile's original request (with reduced quality of service if required), as well as several additional fields such as the IP address of the SGSN.

The GGSN examines the PDP context request, and reduces the QoS if required (if the current load on the GGSN is too high, for example). If the mobile does not have an IP address and the GGSN has been configured to allocate IP addresses itself, then the GGSN allocates a dynamic IP address for the mobile. It then composes a message called *create PDP context response* (6), which it sends to the SGSN.

The SGSN now knows the quality of service that it wishes to implement, so it can allocate a radio access bearer for the PDP context (7), using the same procedures as for a voice call. Once this has finished, the SGSN can respond to the mobile's original request (8). The message includes the QoS that the network is willing to provide, and any IP address that has been allocated by the GGSN. The mobile can now communicate with the server, in the manner described in the next section.

This process has left the mobile in PS-CONNECTED state and RRC Connected mode, and it will continue in these states as long as the data rate is high enough. If the data rate drops, however (for example, if the user pauses to look at a file that he or she has downloaded), then the core network may tear down its signalling connection with the mobile. As part of this step, it tears down the radio access bearer that was set up in step 7, and it places the mobile in PS-IDLE and RRC Idle. The PDP context can stay active, but steps 1, 2, 3 and 7 will have to be repeated if the user tries to communicate with the server again.

In these figures, we have assumed that the mobile is implemented as a single device. If it is divided into a TE and an MT, then the two devices communicate using a low level protocol such as PPP, and extra signalling messages are required between them. We have also assumed that the mobile initiates the process. The network can ask the mobile to set up a PDP context using an SM message called *request PDP context activation*, and the mobile can either reject the request or respond as in step 4.

The mobile may have to exchange application layer signalling messages with the server, for tasks such as authentication and authorisation. These messages can be exchanged in two ways. In *non-transparent access*, the application layer signalling is integrated into the process shown above, so that the mobile specifies the relevant application layer parameters as part of step 4, and the GGSN exchanges signalling messages with the server between steps 5 and 6. In the simpler technique of *transparent access*, the application layer signalling takes place in the UMTS user plane, after all the communications with the GGSN have been established.

A slightly different procedure is required if the mobile sets up a subsequent data transfer from the same APN, with a different QoS profile, while the first one is still active. (For example, the mobile might start to download an audio stream from a server, and then start a separate video download as well.) This procedure is known as *secondary PDP context activation*. In place of step 4 above, the mobile sends an SM message called *activate secondary PDP context request*. As part of this message it specifies a *traffic flow template* (TFT), which describes the IP packets that are associated with the new PDP context. (The template might include the port number used by the layer 4 protocols TCP and UDP, for example.) When an incoming packet arrives, the GGSN compares it with all of the traffic flow templates and sends it to the appropriate PDP context, so that it can receive the correct quality of service on the path to the mobile. Note that there are no such things as primary and secondary PDP contexts: there are just two or more PDP contexts, which are set up using different activation procedures, and which are distinguished using one or more traffic flow templates.

### 5.4.3 Transport of packet data

Figure 5.9 shows the user plane protocol stack, for the case where the PDP type is IP. In the highest layers of the stack, the application software lies above a suitable application layer protocol such as the *wireless application protocol* (WAP). This offers similar services to a normal web browser, but modified to handle the limited input and output capabilities of a mobile device. Data are then transported between the server and the mobile using TCP or UDP over IP.

Inside the core network, data are routed between the GGSN, SGSN and SRNC using the GPRS tunnelling protocol user part (GTP-U), whose operation will be described below. Between the SRNC and the mobile, the packet data convergence protocol (PDCP) compresses the headers of the overlying IP packets, so as to reduce the load on the air interface.

Let's look at the path taken by an IP packet that the server is sending to the mobile. If the mobile has a public IP address, then the server

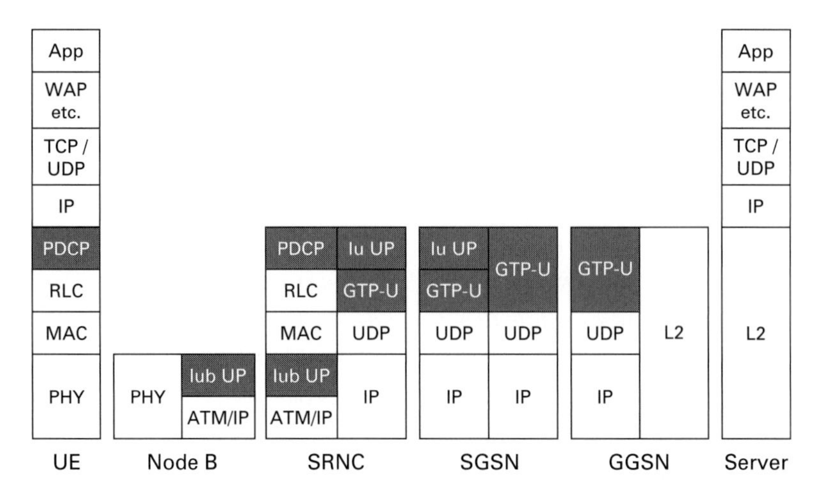


**Figure 5.9** User plane protocol stack used for data flows in GPRS.

addresses the packet using the IP address of the mobile itself. The address belongs to the home network, so the packet arrives at the home network's GGSN. The GGSN then has to send the packet to the correct SGSN, which it does by a mechanism known as tunnelling.

Tunnelling works as follows. Inside the GGSN, the GTP-U protocol examines the mobile's IP address, looks up the PDP context that it should send the packet to, and looks up the IP address of the corresponding SGSN. (If the mobile is associated with more than one PDP context, then the GGSN examines its traffic flow templates as well, so as to identify the correct PDP context for that particular packet.) The GGSN then wraps the incoming packet inside a larger UDP/IP packet, addresses the new packet using the IP address of the destination SGSN, and sends it into the core network. This implies that IP is being used to handle two separate levels of routing: one from the server to the mobile, and one from the GGSN to the SGSN.

The packet arrives at the SGSN, which runs much the same procedure: it unwraps the original packet, identifies the correct SRNC, and sends it there through a second tunnel. Although there is still an Iu user plane protocol between the two network elements, it is placed in a different mode from the speech case that is known as *transparent mode*. As its name implies, this simply passes the packet to and from the underlying transport protocols, without modification.

From release 7, the specifications also support a technique called the *direct tunnel approach*. Using this approach, packets can be tunnelled directly from the GGSN to the SRNC, without the need to pass through the SGSN. This approach cannot be used in all situations (it is unsuitable for roaming mobiles, for example), but if it is used, it reduces the end-to-end delay.

If the mobile's IP address is private to the UMTS network then the process is much the same, except that the core network requires a network address translator, which is typically implemented as part of the GGSN. The server sends its packet to the network address translator, which replaces its own IP address with the private IP address of the mobile. This process is essential for network access through a visited GGSN, to ensure that downlink packets arrive at the visited network rather than the home one.

## 5.5 Other user and bearer services

UMTS supports many other services, from common ones such as SMS and MMS to others that are hardly used at all. It is not possible to give a full description of each service in a book of this size; instead, the next two sections contain an overview of their architecture and operation. We describe the remaining user and bearer services in this section, and cover the UMTS toolkits in Section 5.6.

### 5.5.1 Short message service

The *short message service* (SMS) allows the user to send a short message to another device. The messages have a maximum length of 140 bytes (160 7-bit characters), but individual messages can be concatenated, and they can contain information such as ringtones and low resolution pictures as well as text. The service was introduced into GSM from the mid 1990s and has been phenomenally successful.

SMS is actually implemented as two separate services, mobile originated and mobile terminated. The messages are sent by way of a *service centre* (SC), which stores mobile originated messages and forwards them as mobile terminated messages to their destinations (Figure 5.10). The service centre communicates with the core network through two devices: an *SMS interworking MSC* (SMS-IWMSC) for mobile originated messages, and an *SMS gateway MSC* (SMS-GMSC) for mobile terminated messages.

SMS messages are transported using a separate SMS protocol stack that handles issues such as data formatting and reliable message delivery. In the core network, the messages can be transported across either of the two domains: packet switched transport is usually preferred, unless the user is already taking part in a voice call. In the air interface's access stratum, the messages are transported using control plane signalling, by means of RRC and RANAP direct transfers. (This is another example of cross-over between the user and control planes.) The RRC protocol uses signalling radio bearer (SRB) 4 if it has been set up, or SRB 3 otherwise.

When the mobile sends a message, the SMS protocols deliver it to the service centre and send the mobile an acknowledgement that the message has arrived. The service centre then forwards the message to the

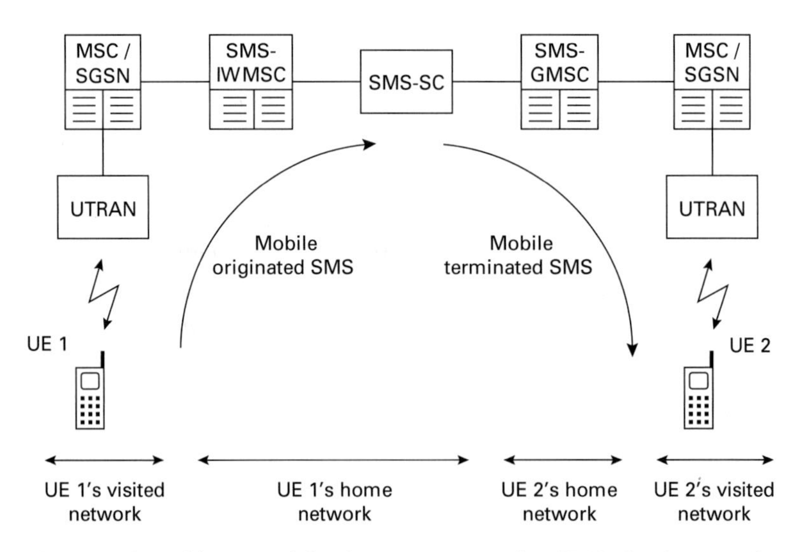


**Figure 5.10** Architecture of the short message service (SMS) for the case of two roaming mobiles. (Adapted from 3GPP TS 23.040.)

destination network's SMS-GMSC. If the mobile is reachable then the SMS-GMSC delivers the message and returns an acknowledgement; otherwise, the SMS-GMSC returns a failure indication, and the service centre tries again later. If the originating mobile is to receive an acknowledgement when the message reaches the destination device, then the service centre has to send it a second mobile terminated SMS.

### 5.5.2 Multimedia messaging service

Although SMS has been extremely successful, it is limited by its short message size. This led to the introduction of the *multimedia messaging service* (MMS) in release 4. This service is the one used to send high resolution picture messages, as well as information such as video files and multimedia.

From a technical perspective, the main difference from SMS is that the service is not fully standardised. Instead, MMS is best thought of as a service framework, which extends up to the application layer and includes support for many common media formats such as JPEG and MPEG-4, but which leaves a lot of flexibility to network operators for its detailed implementation. Crucially, however, the specifications do cover inter-operation between different networks, by including a precise description of how media should be exchanged between them.

Figure 5.11 is an architectural diagram of MMS. The *MMS user agent* is an application layer protocol in the mobile, which sends and receives multimedia messages and informs the MMS relay/server about capabilities such as its screen resolution. The *MMS relay/server* stores multimedia messages and forwards them to the end user, so has similar functions to the SMS service centre. The *user profile database* stores information about the users' service environments, such as the types of message they are willing to receive and any blacklists of users from whom they do not wish to receive messages.

The most important interface is MM1, which lies between the MMS user agent and the MMS relay/server. On this interface, the specifications define a number of *abstract messages*. These specify the requests and responses that the two devices should exchange, together

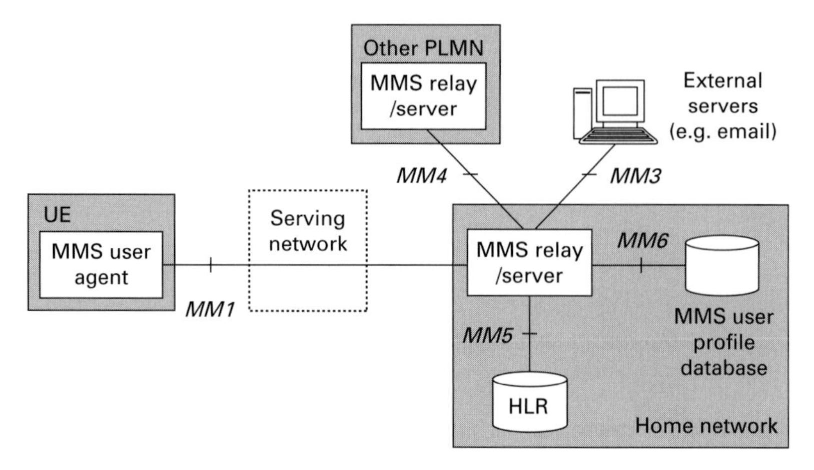


**Figure 5.11** Architecture of the multimedia messaging service (MMS). (Adapted from 3GPP TS 23.140.)

with the associated information elements. (For example, a user agent submits a multimedia message to the relay/server using the abstract message *MM1_submit.REQ*. As part of this message, the user agent specifies information including the recipient's phone number or email address and the content type, and attaches the content.) The abstract messages are implemented using standard protocols such as the wireless application protocol, but the implementation details are up to the network operator.

The message is usually transported in the user plane of the packet switched core network. The MMS relay/server communicates with the PS domain using a GGSN, and is often associated with its own access point name.

The MM4 interface lies between the relay/servers run by different network operators. On this interface, multimedia messages are exchanged using the standard email protocol SMTP. To support inter-operation between different networks, the specifications describe how the MMS information elements are mapped onto SMTP header fields. The MM3 interface connects a relay/server to an external email or web server, so that users who do not have a high resolution display can still retrieve their messages. It is used in a similar way to MM4.

### 5.5.3 Multimedia broadcast/multicast service

Like other mobile telecommunication systems, UMTS was originally designed for point-to-point services such as voice and SMS. Recently, however, there has been increasing interest in broadcast and multicast services such as mobile TV. Delivery of these services over a release 99 network is very inefficient, because the data have to be replicated for every target mobile. The *multimedia broadcast/multicast service* (MBMS) was therefore introduced as part of release 6, to support these services in a more efficient way.

MBMS actually comprises two separate services. The *MBMS bearer service* delivers broadcast and multicast streams over the network in such a way that each stream is only transmitted once per cell, while the *MBMS user service* is the broadcast or multicast application. The two are normally used together but can be kept separate. For example, the MBMS bearer service can be used to transport other applications, while the MBMS user service can be delivered using other mechanisms such as GPRS.

The distinction between broadcast and multicast services is that the former are delivered to all the mobiles in a given broadcast area, while the latter only go to mobiles that have explicitly joined the corresponding multicast group. (A broadcast service might be used for traffic reports, for example, while a multicast service would normally be used for subscription-based mobile TV.) For both types of service, the network can deliver different data streams to different local areas, so that a user can subscribe to a single mobile TV station but receive different local variants in different parts of the country.

The architecture of MBMS is shown in Figure 5.12. The most important new component is the *broadcast/multicast service centre* (BM-SC). This receives multimedia data from a content provider and delivers it over the packet switched core network, and also handles functions such as user subscriptions, service announcements and end-to-end security. The new Gmb interface handles MBMS-specific signalling messages between the BM-SC and the GGSN, while the older Gi interface handles more general messages.

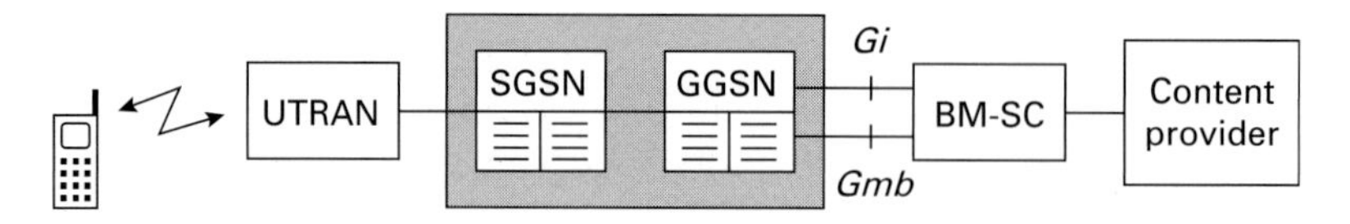


**Figure 5.12** Architecture of the multimedia broadcast/multicast service (MBMS). (Adapted from 3GPP TS 23.246.)

In the core network, MBMS data and signalling messages are transported using IP multicast, while in the radio access network, they are distributed to all the cells containing mobiles that have subscribed to the corresponding service. Within each cell, the information is usually handled using a point-to-multipoint transmission, in which an MBMS bearer is transmitted only once but can be received by all the subscribing mobiles. In this situation, the data are transmitted on the FACH and SCCPCH, with some enhancements to the air interface that are beyond the scope of this book.

Security is provided by application layer authentication, encryption and integrity protection between the mobile and the BM-SC. To ensure that all the target mobiles can decrypt the service, the keys are service specific rather than mobile specific, and are distributed using the secure channel that has already been established between the mobile and the network. The keys are updated regularly, to prevent malicious users from discovering and publishing them.

### 5.5.4 Less common services

UMTS supports a number of other services, which we will briefly describe here.

The *packet switched streaming service* was introduced in UMTS release 4. It delivers real time multimedia streams, such as audio and video, from a streaming server to a mobile over the packet switched core network. It is distinguished from MMS by being a real time service, and from multimedia telephony services by its one-way nature.

The *circuit switched multimedia service* is used for video telephony in UMTS. It is based on the wireless multimedia terminal defined in

ITU specification H.324M: the UMTS specification is often known as 3G-324M. H.324 terminals can support a variety of codecs for audio, video and embedded data: in the case of UMTS terminals, support for AMR voice and H.263 video is mandatory, while MPEG-4 video is optional. The service requires high bit rates: these are typically between 32 and 384 kbps depending on video quality, although the maximum bit rate in release 99 is 64 kbps due to the limitations of the MSC.

The *circuit switched data service* has been part of the system since the days of GSM. It transfers data streams over the circuit switched core network, and was used by the early wireless application protocol (WAP) phones before the introduction of GPRS. It has been more-or-less completely superseded by GPRS, which is far more appropriate for the transport of bursty data streams.

The *cell broadcast service* (CBS) was also carried over from GSM. It allows network operators to broadcast text messages to users on a cell-by-cell basis, for information such as emergency messages and weather reports. The main technical difference from SMS is that the messages are sent to all the users in a cell using a single radio broadcast, instead of requiring multiple messages. Unlike SMS, it has not yet been widely implemented.

## 5.6 Toolkits

Toolkits are low level services, which are not directly visible to the user. Instead, application developers use them to build up higher level services. In this section, we describe the toolkits that are defined in UMTS, and show how some of them come together in a concept known as the virtual home environment (VHE).

### 5.6.1 Location services

Location services allow the network to determine the physical location of a mobile. Possible applications include map-based services for the user, lawful interception by the police or security agencies, and emergency calls.

The case of emergency calls is particularly important. One of the main drivers behind location services has been the wireless E911 mandate of the US Federal Communications Commission, which obliges network operators in the USA to introduce technology that can measure the position of mobiles making emergency calls. When the handset is actively involved in the positioning technique, the required accuracy is 50 m with a probability of 67 per cent, and 150 m with a probability of 95 per cent.

In UMTS, the achievable positioning accuracy is sensitive to the exact technique used. At one extreme, the network can locate a mobile using its cell identity, with an order-of-magnitude accuracy of 1000 m in a macrocell. At the other extreme, a mobile can pinpoint its location to around 10 m using the Global Positioning System (GPS), so long as it can receive a signal from enough of the GPS satellites. If it is surrounded by tall buildings or indoors, then the satellites are often obscured and GPS is unavailable.

The most interesting technique for this book is one called *observed time difference of arrival* (OTDOA). This is a triangulation technique, in which the mobile measures the times at which signals arrive from several Node Bs, and passes the time differences back to the radio access network. The SRNC then estimates the mobile's location by combining the measurements with its knowledge of the Node Bs' locations, and passes the result back to the core network and the application.

There is one difficulty with OTDOA: if a mobile is too close to one base station, then it will be unable to hear any others, and will be unable to make any timing measurements. This is overcome by the use of *idle periods in the downlink* (IPDLs), in which individual Node Bs stop transmitting for a short time period (either ½ slot or 1 slot), so that mobiles can hear Node Bs that are further away. In some ways, idle periods are similar to transmission gaps in compressed mode, but with a crucial difference: a transmission gap only affects one mobile at a time, while an idle period affects all the downlink transmissions from a Node B.

In principle, the achievable accuracy using OTDOA is of the order of 80 m (roughly the distance travelled by radio waves over the duration of one chip). This can be degraded by multipath, although intelligent processing algorithms can recover some of the loss. The result is an accuracy that is tantalisingly close to the requirements of E911.

### 5.6.2 CAMEL

Intelligent networking is a technique widely used by fixed network operators to provide value added services such as prepay and freephone numbers. CAMEL (*customised applications for mobile network enhanced logic*) is an extension of intelligent networking for GSM and UMTS that supports roaming. It is implemented in such a way that roaming users continue to receive intelligent networking services from their home network, in the same way that they do if they are not roaming. CAMEL is applied not only to voice calls, but also to services such as SMS and GPRS.

Figure 5.13 shows the architecture of CAMEL in the circuit switched domain. The CAMEL services are implemented in the *GSM service control function* (gsmSCF), which is located in the home network. The services are invoked by messages received from the *GSM service switching function* (gsmSSF), which is associated with the visited network's MSC.

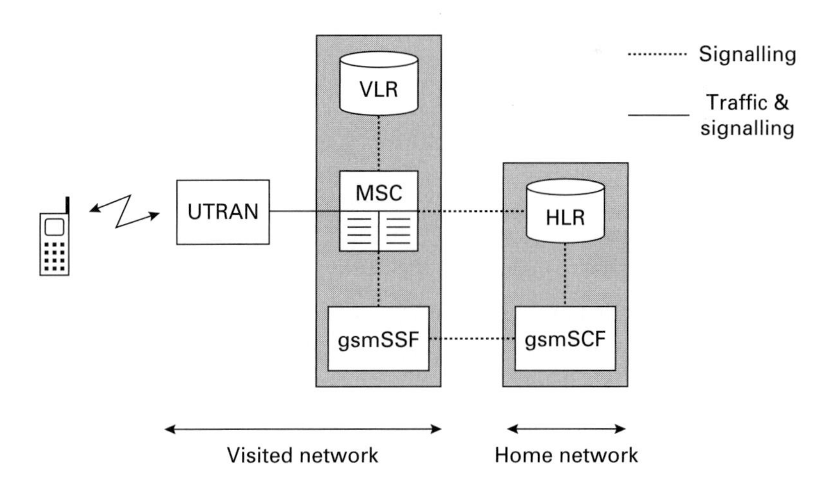


**Figure 5.13** Architecture of CAMEL for a roaming mobile. (Adapted from 3GPP TS 23.078.)

The gsmSSF and gsmSCF communicate using an SS7 application layer protocol known as the *CAMEL application part* (CAP).

One use of CAMEL is in number translation. The user might dial a shortcut number, such as a number provided by the home network operator for topping up a prepay phone. In step 4 of Figure 5.2, the mobile's call control protocol sends a setup message to the MSC containing the shortcut number. This causes the gsmSSF to send a CAP message to the home network's gsmSCF, in which it includes the shortcut. The gsmSCF replies with the underlying telephone number, and the MSC can continue to process the call as before.

### 5.6.3 USIM application toolkit

The *USIM application toolkit* (USAT) is a protocol that runs inside the mobile, between the UICC and the ME. It provides the two devices with a set of commands that allow applications to run on the UICC and control the behaviour of the ME. The protocol interactions are shown in Figure 5.14.

The commands fall into two main categories. Some of them allow a UICC application to control interactions with the user, for example by telling the ME to display a menu and return the user's response. Others allow the application to control interactions with an external server, for example by telling the ME to send an SMS message to the

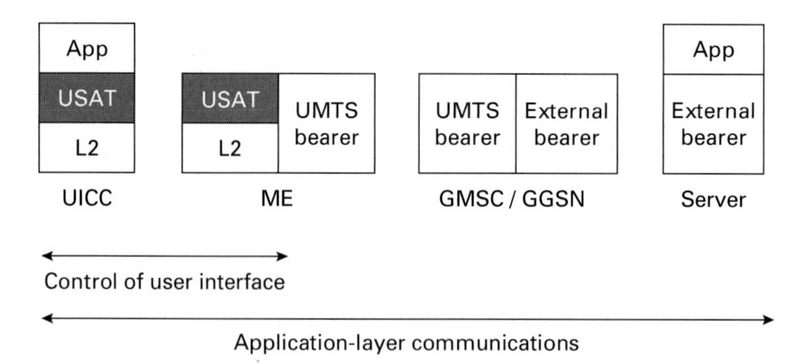


**Figure 5.14** Protocol interactions for the USIM application toolkit (USAT).

server and await the reply. Together, they allow an application on the UICC to take control of the behaviour of the mobile.

Two other features of USAT are worth noting. First, UICC applications can be either pre-installed by the network operator, or subsequently downloaded by the user. Second, the specification includes security procedures, which ensure that any SMS messages are securely transferred from the sending to the receiving network.

A typical use of USAT is in e-commerce. A user might download a simple billing application from a utility operator. The application asks the user for the numbers of his or her utility and bank accounts, and sends the information by SMS to the utility operator, where it is used to pay the bill. This could be done using a web server, of course, but the use of USAT has a few advantages: the application stays with the user when he/she upgrades to another ME, the SMS data are transmitted securely, and the communication requirements are reduced because most of the process takes place locally.

### 5.6.4 Mobile execution environment

The capabilities of UMTS mobiles can vary enormously from one device to the next. In Chapter 2, we saw that the SRNC could find out a mobile's low level capabilities, such as its maximum data rate, by sending it a capability enquiry message. However, that process does not help with higher level capabilities, that are of no interest to the SRNC. For example, some devices have a low screen size and can show very little information at one time, while on others the screen size is much higher.

This gives service providers a problem: how do they find out the high level capabilities of a mobile, so that they can supply it with suitable content? The answer lies in the *mobile execution environment* (MExE).

The basic architecture of MExE is shown in Figure 5.15. Using a standardised technique known as *composite capability/preference profiles* (CC/PP), the ME tells a server that supports MExE about its capabilities. (The CC/PP messages are typically transported using GPRS.) The most important capability is the MExE classmark, which describes

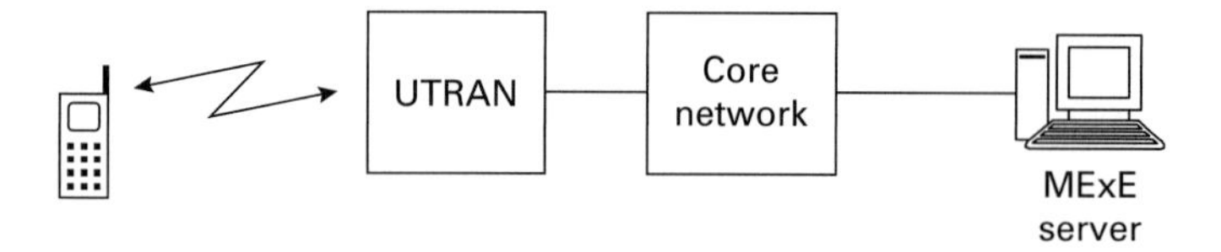


**Figure 5.15** Architecture of the mobile execution environment (MExE). (Adapted from 3GPP TS 23.057.)

the application environment installed in the ME, such as WAP or PersonalJava. Other capabilities include the ME's screen size, support of colour and type of keyboard. The ME can also specify content preferences to the server, such as the character set to use and the preferred language. The server then uses the information to supply content which is formatted in the requested way and which is suitable for the ME.

A good example is the use of MExE to support web browsing. Once the server knows the ME's capabilities and the user's content preferences, it can supply it with web pages that are dimensioned according to its screen size and written in the requested language. However, the uses of MExE do not stop there. For example, it can also be used to support the downloading of applications such as web browsing clients, by delivering software to the ME that is appropriate for its application environment and its processing and storage capabilities.

### 5.6.5 Open service access

Traditionally, telecommunication services have been provided by the network operator and have been closely coupled with the operation of the network. *Open service access* (OSA) allows applications to communicate with the network across an open, standardised interface, the OSA application programming interface (API). This allows third-party developers to write applications that are independent of the underlying network technology.

The basic OSA architecture is shown in Figure 5.16. The network contains a number of *service capability servers* (SCSs), which communicate with an OSA application across the OSA API. The SCSs provide

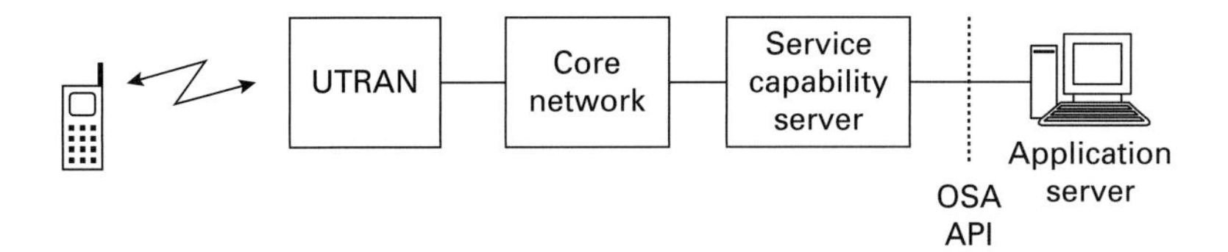


**Figure 5.16** Architecture of open service access (OSA). (Adapted from 3GPP TS 23.198.)

the application with *service capability features* such as call control, messaging, location measurement and charging. The application can use these features without knowing anything about how they are implemented in the network.

As a simple example, a user might send an SMS message to an OSA application server, to ask where he or she is. The OSA application determines the user's position by means of the location service described above, using the service capability features provided by the OSA API. It then sends the user an MMS message to show that position on a map. At the same time, the application uses the charging system described below to debit the user's account and to invoice the network operator.

### 5.6.6 Virtual home environment

At first glance, there are few similarities between CAMEL, USAT, MExE and OSA. Collectively, however, they form an important concept known as the *virtual home environment* (VHE).

The aim of the VHE is that a user should be presented with the same services, personalised service features and user interface customisation, irrespective of the network that he/she is roaming in, or the terminal that he/she is using. Thus, CAMEL, USAT and OSA allow applications to run in the same way on different networks, while MExE does the same job for different terminals.

To support the VHE, an application can associate each user with a *user profile*, which describes the way in which the user would like to receive services, and which is stored in the UICC and/or the home network. For example, MExE user profiles are stored in the UICC and specify the user's content preferences to MExE applications. This ensures that a user can (for example) still receive web pages from an MExE server in the preferred language, even after upgrading to another ME.

Clearly this can result in a proliferation of user profiles, so 3GPP are developing a concept known as the *generic user profile* (GUP). This is a user profile which is written in a consistent format and which can be made available to any application at all.

## 5.7 Charging and billing

We close this chapter with a brief discussion of charging and billing, which are closely related to service provision. In discussing them, it is best to begin with a few words about the terminology used.

A *chargeable event* is an activity that the user might be charged for, such as the beginning or end of a call, or the delivery of an SMS message. A *charging data record* (CDR) is a collection of information about a chargeable event, which is written in a standardised format. *Charging* is the process of collecting CDRs and sending them to a billing system, while *billing* is the process that transforms the CDRs into bills requiring payment. Our main interest in this section is in charging: billing lies outside the 3GPP specifications and is an issue for the network operator alone.

There are two varieties of charging. Contract phones use *offline charging*. The charging system sends the CDRs to a billing system, which works out the bill and sends it to the user. The charging information has no effect on the real time service. On the other hand, prepay phones use *online charging*. Using this technique, the charging information does affect the real time service: the charging system authorises the call before it begins, debits the user's account in real time, and may cut off the call if the user runs out of credit. We discuss these techniques one-by-one below.

There is also a distinction between event-based and session-based charging. Services like SMS and MMS use *event-based charging*, which is done on the basis of a single event (e.g. a single message transfer). Voice and GPRS are more likely to use *session-based charging*, which is done from several events that collectively describe the call duration and the amount of data transferred.

### 5.7.1 Offline charging

Figure 5.17 shows the architecture of the offline charging system. The network nodes each contain a *charging trigger function* (CTF), which generates information about chargeable events. The *charging data function* (CDF) uses this information to write the charging data records, and the *charging gateway function* (CGF) collates the CDRs and sends them to the billing system. The CDF and CGF can be integrated into the network nodes in the same way as the CTF, or they can be separate.

If a subscriber is roaming, then the home and visited networks both collect charging data records. The visited network then uses its CDRs in two ways: it submits them to the home network as part of the *transferred account procedure*, so that the home network can bill the subscriber, and it uses them to invoice the home network for the subscriber's use of the network's resources.

### 5.7.2 Online charging

Figure 5.18 shows the architecture of the *online charging system* (OCS). In online charging, the CTF generates information about chargeable events as before, but it must also get permission for the requested resource usage from the *online charging function* (OCF). The *rating function* (RF) determines how much the user's account should be debited, and the *account balance management function* (ABMF) informs the OCF about the user's balance and debits it on request.

The OCS can grant permission for resource usage in two ways. For event-based charging such as SMS message transfer, the OCS debits

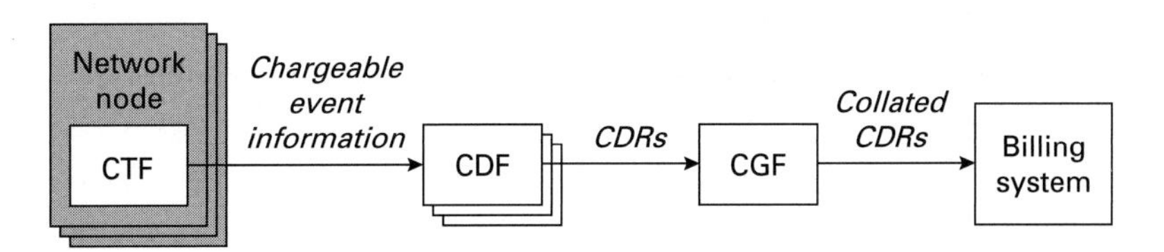


**Figure 5.17** Architecture of the offline charging system. (Adapted from 3GPP TS 32.240.)

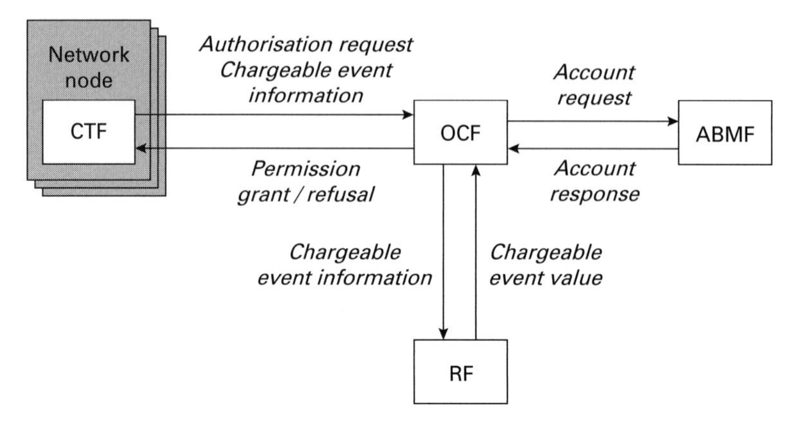


**Figure 5.18** Architecture of the online charging system. (Adapted from 3GPP TS 32.240.)

the user's account and grants permission to the CTF, before the resources are used. The OCS can also make a reservation in the user's account until the event has completed, in case the requested event fails. For session-based charging, the OCS grants permission for a limited amount of resource usage such as a ten-minute voice call, and makes a reservation in the user's account as before. If that permission expires, then the CTF requests permission for more resources, which can be granted or withheld as required. On completion of the call, the CTF tells the OCS about the actual resources used, and the OCS debits the user's account.

If the subscriber is roaming, then the OCS is in the home network, and the home and visited networks both contact it to debit the subscriber's account. At the same time, the visited network collects CDRs, and uses them to invoice the home network for the subscriber's use of resources. In the CS and PS domains, online charging is implemented using CAMEL, so it makes immediate use of the roaming support that CAMEL provides.

# 6 Future developments

In the final chapter, we will look at some of the likely future developments of UMTS. We begin with the IP multimedia subsystem, which was introduced into the 3GPP specifications in release 5, and is intended for the delivery of real time, packet switched services to the user. We continue with a look at the long term evolution of UMTS, which is intended to be part of release 8, and will involve changes to the radio interface that will supply the user with much higher data rates than before. We conclude with an overview of the process for defining fourth generation systems, and a look at some of the likely candidates.

## 6.1 The IP multimedia subsystem

The *IP multimedia subsystem* (IMS) is an extra component of the fixed network. Its main objective is to deliver real time services such as voice and video over the core network's packet switched domain, which have not been supported by previous implementations of UMTS. This section describes the objectives and architecture of the IP multimedia subsystem. It then gives an overview of the protocols and operational procedures that it uses, and describes the services that have been defined for use on the IMS.

### 6.1.1 Objectives

The IMS is intended to bring three main benefits to network operators and to the user.

First, the IMS provides good end-to-end quality of service for packet switched data streams. Until now, the network has provided a guaranteed bit rate and a low delay over the path between the mobile and the GGSN, but has made no such guarantee beyond the GGSN. The IMS can monitor and control the quality of service in certain kinds of external network such as the interconnecting backbones between two GGSNs, for the benefit of applications such as voice over IP. This should eventually allow network operators to handle all voice and multimedia calls over the packet switched domain and, ultimately, to retire the circuit switched domain altogether.

Second, the IMS has an open application architecture, which makes it easy for third-party developers to write applications for it. The IMS will provide a number of common application support functions such as presence services and group list management, so that application developers will not have to write these for themselves.

Third, users will be able to reach the IMS using not only a UMTS network, but also other access technologies such as wireless local area networks. This will benefit the user, who will be able to reach exactly the same services using a mobile phone, a laptop or any other device. It will also benefit the network operators, as by investing in the IMS, they will be protected against the risk that some other access technology will eventually prove more popular than 3G.

The first IMS specifications were part of UMTS release 5 and have attracted considerable interest. (For example, 3GPP2 have designed an IMS for use by cdma2000 networks, and have based it very closely on the one designed by 3GPP.) However, network operators have been slow to introduce the IMS, for two main reasons: it requires a lot of investment and its benefits will take a long time to realise.

### 6.1.2 Architecture

Figure 6.1 is an architectural diagram of the IMS. Dashed lines denote traffic while dotted lines denote IMS signalling. As implied by the figure, most of the richness of the IMS lies in the signalling messages that set up and control the traffic streams. The IMS is more-or-less independent of the core network, so in the discussion that follows, we will see that the IMS contains close analogues to many of the earlier core network functions.

The most important components of the IMS are the *call session control functions* (CSCFs). There are three types of these. The *serving*

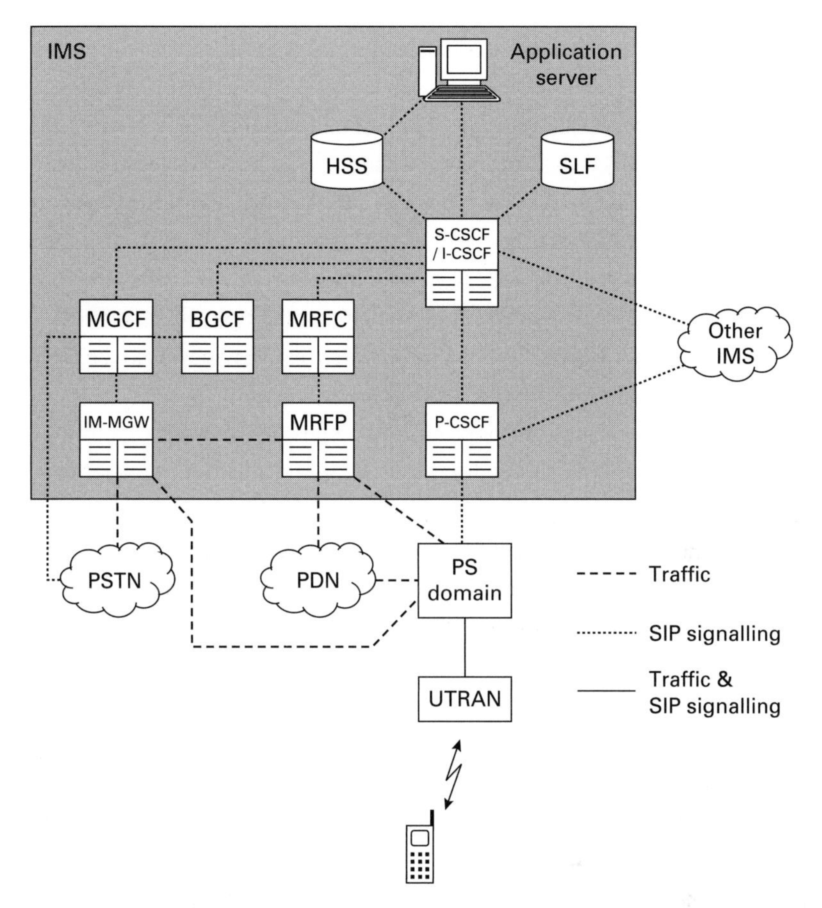


**Figure 6.1** Simplified architecture of the IP multimedia subsystem. (Adapted from 3GPP TS 23.002.)

*CSCF* (S-CSCF) is the mobile's main point of contact in the IMS. It has functions similar to the signalling functions of the MSC or the SGSN, and carries out tasks such as location registration and call control. The *proxy CSCF* (P-CSCF) is an interface between the IMS and the packet switched core network. It compresses signalling messages on the path between the IMS and the mobile, and applies encryption and integrity protection to them. It also manages the user's quality of service, by instructing the GGSN to make resource reservations and monitor the traffic flows. The *interrogating CSCF* (I-CSCF) is an interface into the IMS, for signalling messages that arrive from the outside world.

Two components manage the subscriber database. The home subscriber server (HSS) has the same functions as in the release 5 core network. The HSSs in the two networks can be either the same physical device or different ones, to handle the case where mobiles use one operator's core network to access another operator's IMS. In the IMS, the HSS's functionality can be distributed over different physical devices: if this is done, then the *subscriber location function* (SLF) points to the HSS where a particular subscriber's details can be found.

*Application servers* provide various types of service. Examples include support functions such as presence and group list management, services standardised by 3GPP such as push-to-talk over cellular (PoC), and other services provided by the network operator or third-party developers. Interfaces are also available to the CAMEL and OSA servers that we saw in Chapter 5, to ensure that services already written for the CS or PS core networks can still be provided over the IMS.

Most of the other components relate to traffic. The *IMS media gateway* (IM-MGW) is an interface to an external CS network, so it allows a mobile to communicate with a traditional telephony system over the IMS. The *media gateway control function* (MGCF) carries out low level control of the IM-MGW, and is itself controlled by the S-CSCF or by an application server. Related to these is the *border gateway control function* (BGCF), which selects the correct media gateway to use in the case of an outgoing call. The *media resource function processor* (MRFP) mixes IP multimedia streams for applications such as conference calls, and is controlled by the *media resource function controller* (MRFC).

It is worth saying a few words about the architecture for a roaming mobile. The S-CSCF is always in the subscriber's home network, and all signalling messages are routed back there. The P-CSCF and the GGSN are always in the same network as each other, but this can be either the home network or the visited one. It is better to have them both in the visited network, as the traffic does not have to go back to the home network, but this solution can only be used if the visited network has an

IMS of its own. If the visited IMS has not been built yet, then the P-CSCF and GGSN must both go in the home network.

Each IMS user has one private identity and one or more public identities. The private identity is analogous to the IMSI, and identifies the user to the IMS. Public identities identify users to each other, and have two main forms: an identity such as sip:user@operator.com looks like an email address, while an identity such as tel: +44-1223-123456 looks like a phone number. Both types of identity are stored on the UICC, in the *IP multimedia services identity module* (ISIM).

The 3GPP IMS is designed around IP version 6. Early implementations of the fixed network are allowed to use IPv4, but this flexibility is not available for the mobile, for which IPv4 is optional (for communication with those early networks) but IPv6 is mandatory. The situation is different in the 3GPP2 IMS, which supports IPv6 and IPv4 equally.

### 6.1.3 Protocols

The main signalling protocol in the IMS is the *session initiation protocol* (SIP). SIP is defined by the *Internet Engineering Task Force* (IETF): it is based on the familiar hypertext transfer protocol (HTTP), and is already widely used by fixed line voice over IP systems. SIP is implemented in both the IMS and the mobile, and its tasks include location management, call control, and internal signalling in the IMS.

In a departure from previous telecommunication protocols, SIP is text based, which makes the messages more readable than before but also larger. Each SIP message has three parts. The first line is either a text request, or a numerical response with an associated text description. For example, a mobile might send a SIP message called *INVITE* to ask another mobile to take part in a voice or video call, and the second mobile might accept the invitation using a positive response called *200 OK*. There are then several header lines, which handle tasks such as routing. Finally, there may be some embedded message content written in another protocol. We will see an example of this in a moment.

The syntax of individual SIP messages is defined by the relevant IETF specifications. However, there are additional 3GPP specifications, which contain information such as the message sequences to be used in the IMS and the precise header lines that are required.

The *session description protocol* (SDP) is used to negotiate the characteristics of media streams, such as the data rates and codecs to be used. In the example above, the originating mobile might describe the requested media and supported codecs in an SDP message, which is delivered by embedding it into the SIP INVITE. The terminating mobile might then modify the SDP message to remove the codecs that it does not support, and embed the information into its 200 OK response.

Traffic is carried by IP and UDP. A third protocol, the *real time protocol* (RTP), supports the delivery of real time packet data, for example by adding timestamps to packets in the transmitting device so that the receiver can play them back at the correct time intervals.

Within the PS core network, IMS traffic and signalling are both treated as traffic, and are transported using packet data protocol (PDP) contexts in the manner shown in Figure 5.9. Usually, one PDP context is used to carry IMS signalling messages, and additional PDP contexts are used for traffic.

### 6.1.4 Procedures

To illustrate how the IMS works, Figure 6.2 shows the procedure for registration. This is analogous to the earlier procedure by which the mobile attaches to the core network, but takes place later and independently. The figure is simplified, in that the full sequence has about 20 step-by-step SIP messages: we have only shown the resultant end-to-end flows. In the description below, we have assumed that the mobile is roaming, and that the GGSN and P-CSCF are in the visited network.

Before the procedure begins, the mobile sets up an RRC connection (Figure 4.6), attaches to the PS core network (Figure 4.7) and sets up a PDP context that will carry the SIP signalling (Figure 5.8). It then sends the IMS a SIP request called *REGISTER* (1), which contains the user's private identity and identifies the user's home IMS. The message arrives at the P-CSCF, which identifies an I-CSCF in the home network by a domain name lookup and sends the message there. On receiving

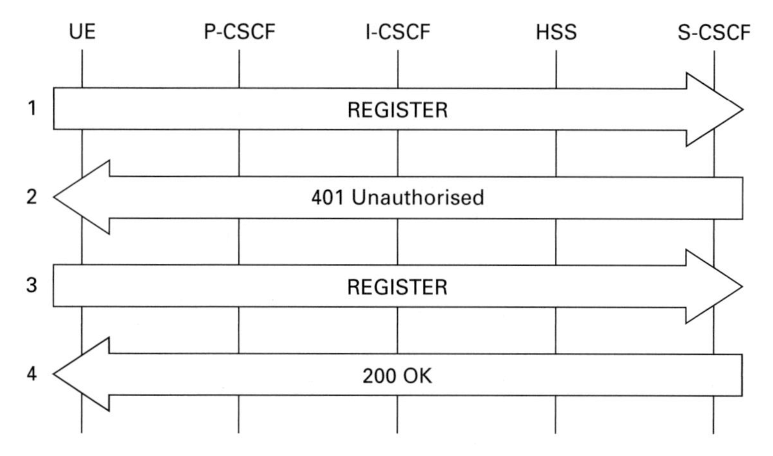


**Figure 6.2** Simplified version of the message sequence chart that a mobile uses to register with the IP multimedia subsystem.

the message, the I-CSCF asks the HSS to identify an S-CSCF that can serve the mobile, and forwards the message to it.

The S-CSCF retrieves an authentication vector for the user from the HSS. The vector is constructed using the same algorithm that was used in the core network (Figure 4.8), but it is otherwise independent of the core network's authentication data. The S-CSCF then sends the mobile a SIP response called *401 Unauthorised* (2), which contains the authentication challenge. This is routed back to the visited network's P-CSCF, and from there to the mobile.

The mobile verifies the identity of the IMS, and responds to the network's challenge with a second SIP REGISTER request (3). The network checks the information that the mobile has supplied, and confirms its registration with a 200 OK response (4).

The mobile now has to keep open the PDP context that it has been using for SIP signalling, until it de-registers from the IMS. This is a departure from previous uses of PDP contexts, which have been associated with individual data flows and have only been temporary.

Once the mobile has registered with the IMS, it can send or receive a call. This involves the exchange of SIP signalling messages between the source and destination mobiles, by way of the IP multimedia subsystems they are registered with. An important part of the call setup process is the negotiation of the media to use. In its initial SIP request, the originating mobile embeds an SDP message that describes the requested media and data rates, and the codecs that it supports. Along the way, this message is examined by several of the network nodes. The S-CSCFs check whether the requested media are compatible with the users' service agreements, the P-CSCFs check whether the networks can handle the requested data rates, and the destination mobile checks whether it supports the requested media and codecs. Any problems are handled by modifying the SDP information, which is returned to the originating mobile. On receiving this information, the mobile knows which media and codecs are acceptable to the other network nodes, and can choose and signal the one it prefers.

### 6.1.5 Services

3GPP have produced specifications for a few IMS-related services, but the slow take-up of the IMS has made this process slower than it might otherwise have been.

The *presence service* allows a user to submit presence-related information to the IMS by means of SIP messages, which is then made available to other users or services that are known as watchers. Presence information describes whether a user is registered with the IMS, where he/she is located and whether he/she is willing to be reached, so it can be used in support of interactive applications such as chatrooms.

The *IMS messaging service* comes in two main varieties: page mode and session mode. *Page mode* messaging is similar to the multimedia messaging service: a user can send messages such as text, pictures or video as content embedded within a SIP request. *Session mode* messaging requires the user to set up an IP multimedia session before any messages can be exchanged, but is more powerful once that has been done: for example, the user can send and receive several messages in a single session, for applications such as chatrooms.

The *IMS telephony service* handles PS conversational multimedia. It uses the call setup procedures that were outlined earlier, but includes additional features such as a minimal set of codecs and procedures for handling packet loss and jitter. It also defines a set of supplementary services that are almost identical to those for CS voice calls.

*Push-to-talk over cellular* (PoC) is a service similar to a walkie-talkie. Users set up an interactive group call, in which one user at a time gets permission to speak by pressing a button on the mobile. Proprietary solutions for PoC have been available for some time (for example, it was an integral part of Motorola's iDEN system), but they have not supported inter-operation between different networks. To deal with this problem, the *Open Mobile Alliance* (OMA) has produced specifications for a fully open PoC service, which is independent of the underlying architecture but which has the IMS in mind. As of January 2008, however, 3GPP have yet to produce any specifications for an implementation of PoC.

## 6.2 Long Term Evolution

The aim of UMTS *Long Term Evolution* (LTE) and *System Architecture Evolution* (SAE) is to keep UMTS competitive with other communication technologies over the next ten years and beyond. It will do this by increasing the data rate to a maximum of 100 Mbps on the downlink and 50 Mbps on the uplink, while reducing both the end-to-end delay and the complexity of the system.

Long Term Evolution addresses the air interface and the radio access network, while System Architecture Evolution is concerned with the core network. At the time of writing, they were important work items in 3GPP, and were intended to enter the specifications in UMTS release 8.

### 6.2.1 Architecture

LTE and SAE are being optimised for packet switched communications, to the extent that circuit switched transport is no longer supported, and all voice calls are carried using voice over IP. There are therefore two main components, the *evolved UTRAN* (E-UTRAN) and the *evolved packet*

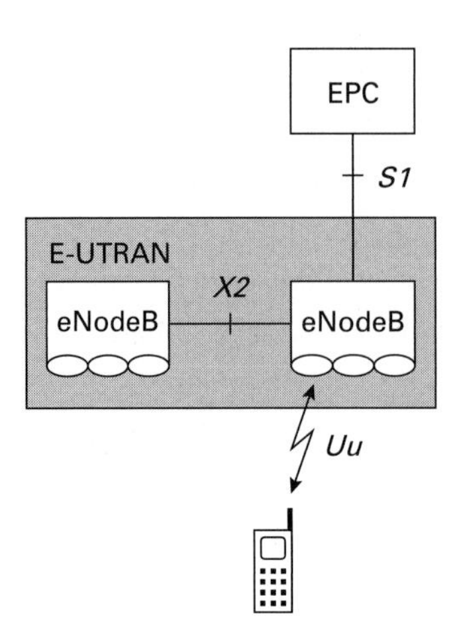


**Figure 6.3** Architecture of the evolved UTRAN. (Adapted from 3GPP TS 36.300.)

*core* (EPC). The EPC is the subject of the System Architecture Evolution initiative, and is still at an early stage of design. The architecture of the E-UTRAN is shown in Figure 6.3.

The E-UTRAN just has one component, the *evolved Node B* (eNodeB). This combines the functions previously associated with the Node B and the RNC, to reduce the system complexity and the end-to-end delay. Soft handover is no longer supported, as it was felt that, in the case of LTE, any performance benefits were outweighed by the extra complexity that it introduced into the mobile. There is still an interface between the eNodeBs, however, for tasks such as signalling and packet forwarding in the case of hard handover.

The air interface no longer uses CDMA: instead, the multiple access schemes are *orthogonal frequency division multiple access* (OFDMA) in the downlink, and *single carrier frequency division multiple access* (SC-FDMA) in the uplink. These perform better than CDMA in the case of high data rate communications. They were not implemented in earlier mobile telecommunication systems for two main reasons: the signal power can vary a lot, which means that the transmitter's power amplifier has to be very accurately linear, and their benefits only appear at high data rates. OFDMA is already used in other systems such as wireless LAN and WiMAX, but is relatively new to mobile cellular networks.

The bandwidth is no longer fixed: instead, LTE supports bandwidths from a minimum of 1.25 MHz (for compatibility with cdmaOne and cdma2000) to a maximum of 20 MHz (where the highest data rates can be achieved). The air interface does, however, reuse several other features that were previously developed for UMTS, such as hybrid ARQ with soft combining, and multiple input multiple output antennas.

### 6.2.2 OFDMA downlink

OFDMA is very like FDMA, but with the crucial difference that the frequency division is done digitally, using a *fast Fourier transform* (FFT). This gives it two main advantages over analogue FDMA. First, it ensures that the individual frequencies are orthogonal, so they can be spaced much more closely together than before without any possibility of interference between them. Second, it allows the network to allocate frequencies dynamically to individual mobiles, in response to their varying data requirements.

OFDMA also has some advantages over CDMA for high data rate communications. In a CDMA system, a high bit rate requires the use of a high chip rate, and hence a low chip duration. This implies that the rake receiver will pick up a large number of distinct rays, so it will need a large number of rake fingers to process them. This can make the receiver extremely complex. In contrast, OFDMA does away with the use of chips, and accepts that the received signal will undergo fading. Instead, it uses more powerful error correction coding techniques that disperse the information in the frequency domain as well as in the time domain, to correct the bit errors that result.

Figure 6.4 shows the basic principles of the eNodeB's OFDMA transmitter. In the figure, the eNodeB wants to transmit eight data streams in parallel, either for a single mobile or for up to eight different mobiles. The eNodeB grabs one symbol at a time from each data stream, and interprets the information as a frequency domain representation of the waveform that it wants to transmit. By carrying out an inverse FFT, it transforms this representation into a block of data in the

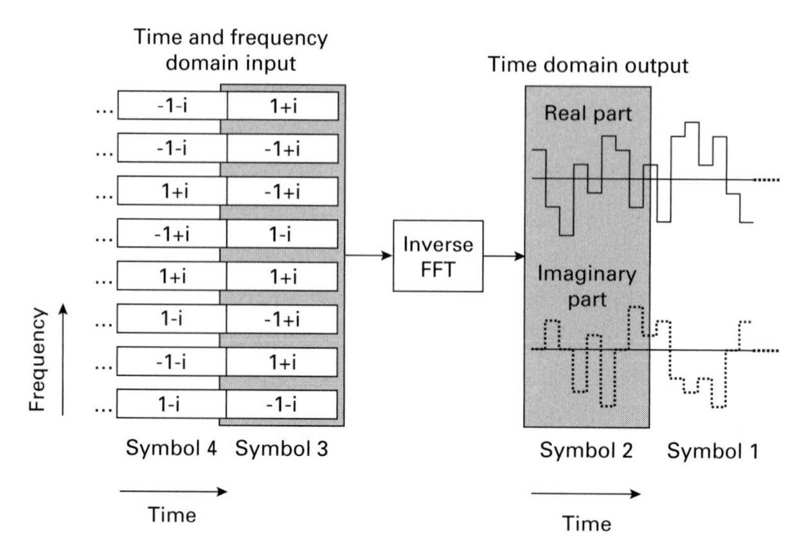


**Figure 6.4** Generation of the transmitted waveform in OFDMA.

time domain. The effect is that each data stream is transmitted on a different frequency, which is known as a *sub-carrier*. The symbol duration can be much longer than in other multiple access techniques, because the eNodeB is transmitting a large number of sub-carriers in parallel.

In a multipath environment, the mobile receives several copies of the transmitted waveform, with different arrival times. This makes successive blocks blur into each other, in a phenomenon known as inter-symbol interference. This can occur using any multiple access technique, but is particularly easy to cure when using OFDMA. We start by leaving gaps between the symbols in each sub-carrier. The gaps are longer than the multipath time dispersion, which prevents successive blocks from overlapping in the receiver. However, they are also much shorter than the OFDMA symbols, so they do not have much effect on the transmitted waveform. After the transmitter's inverse FFT, we fill the gaps in a step called *cyclic prefix insertion*, by copying information from the end of each block into the previous gap. We use a cyclic prefix instead of a gap, because it gives the best results when the data are processed using an FFT.

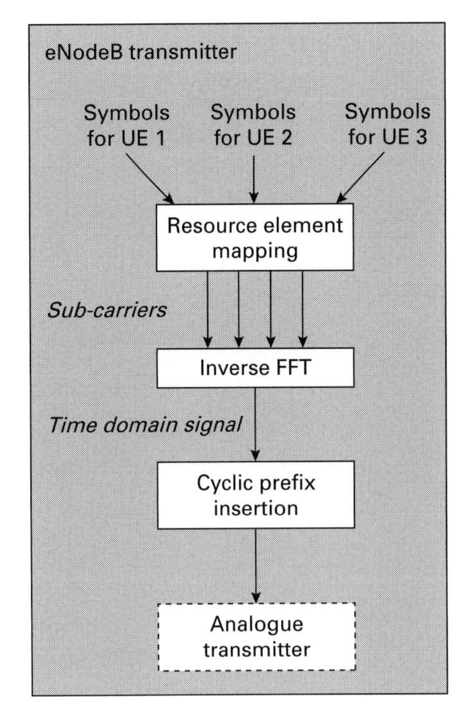


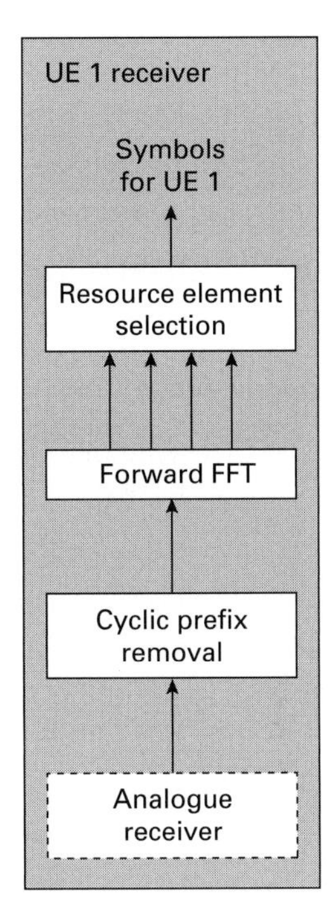


**Figure 6.5** Block diagram of the OFDMA transmitter and receiver in the downlink of the evolved UTRAN. (Adapted from 3GPP TR 25.814.)

The resultant process is shown in Figure 6.5. In the eNodeB, the input is a stream of symbols for each of the mobiles in the cell. The *resource element mapper* maps these streams onto the available sub-carriers. The eNodeB takes one symbol at a time from each sub-carrier, transforms them into a block of time domain data as before, inserts the cyclic prefixes between successive blocks, and sends them to the analogue processor for transmission. The mobile samples the incoming data stream, divides it into blocks, and removes the cyclic prefix from each block. It then passes each block to a forward FFT, which converts the data stream to a set of sub-carriers in the frequency domain. Finally, the mobile selects the sub-carriers that it wants, throws away the remainder, and passes the resulting symbols to higher layers.

In LTE, each sub-carrier usually has a bandwidth of 15 kHz. The properties of the FFT mean that each symbol has a duration of 1/15 ms, but the use of cyclic prefixes makes the interval between successive symbols slightly longer, with an average value of 1/14 ms. The eNodeB allocates its sub-carriers to individual mobiles in units of a *resource block*, which has a bandwidth of 180 kHz (usually 12 sub-carriers) and a duration of 0.5 ms (usually 7 symbols).

### 6.2.3 SC-FDMA uplink

The OFDMA scheme described above has one disadvantage. With one data stream per sub-carrier, it turns out that the power of the transmitted signal can vary a lot. This in turn means that the transmitter's power amplifier has to be very accurately linear, as any non-linearities will distort the waveform. Amplifiers like this are expensive and power inefficient: this is not usually a problem for the base station, but it is a problem for the mobile.

To deal with the problem, the uplink uses a modified version of OFDMA, known as SC-FDMA. The only difference is the addition of a direct FFT before the mobile's resource element mapper. This allows the mobile to disperse each transmitted symbol across all the sub-carriers it is using, so that each sub-carrier contains data not from one symbol, but from several. This reduces the variations in transmit power, and allows the mobile to use cheaper amplifiers than it otherwise would.

## 6.3 Towards 4G

Mobile telecommunication technology has made enormous progress since the first standards for 3G systems were published in 1999. The International Telecommunication Union (ITU) has therefore begun the task of defining what a 4G system will be, under the name *IMT-Advanced*.

The first important document was ITU recommendation M.1645, published in 2003. This is a vision for the future development of 3G systems and for the introduction of 4G. As part of this vision, the ITU proposed that 4G systems should deliver peak data rates up to about 100 Mbps for fast moving users, and 1 Gbps for stationary or slow moving ones.

To deliver these high data rates, 4G systems will require a new allocation of radio spectrum. In 2007, the ITU World Radiocommunication Conference identified a number of new bands for use by mobile cellular networks. Notable among these was a band from 3.4 to 3.6 GHz, which is intended for use by IMT-Advanced in regions of the world outside North and South America.

The ITU intends to publish the requirements for IMT-Advanced in 2008, and will follow this with a call for candidate systems and an assessment of them. The result of this process will be a family of 4G systems, analogous to those for 3G. It is worth stressing that the definition of 4G technologies is still at an early stage, and until the ITU process has completed, it would be incorrect for anyone to describe their system as a 4G one. However, we can identify some of the likely candidates.

UMTS Long Term Evolution will undoubtedly become a candidate technology for 4G, either directly or through some later enhancement. *Ultra Mobile Broadband* (UMB) is a similar project to enhance the performance of cdma2000, which uses many of the same techniques such as OFDMA, variable bandwidth transmission and MIMO antennas. The WiMAX submission for 4G is likely to be based on a new version of the standard, IEEE 802.16m, which is under development at the time of writing.

# Bibliography

In a book of this size, it is impossible to provide anything more than a brief introduction to UMTS. The following books are well written and more detailed than this one, and are all recommended as sources for further reading.

Lescuyer's book is probably the closest to this one, although it has more about the air interface and less about the operation of the network. Unfortunately the English edition only covers release 99, so is now rather out of date. The books by Korhonen and by Bannister *et al.* both cover more material: the first concentrates mainly on the air interface and is exclusively about UMTS, while the second looks more at the network side and also covers GSM.

Engineers working on the air interface are well served by the next four books. Richardson describes each of the air interface protocols to a similar level of detail, while Tanner and Woodard concentrate more on the physical layer. Johnson also covers the signalling procedures inside the radio access network. Dahlman *et al.* look at the later releases of UMTS, by describing high speed packet access and Long Term Evolution.

The remaining books are more specialised ones. Chevallier *et al.* cover the techniques used for network planning and operation in UMTS, and Camarillo and Garcia-Martin describe the IP multimedia subsystem.

J. Bannister, P. Mather & S. Coope, *Convergence Technologies for 3G Networks: IP, UMTS, EGPRS and ATM* (Wiley, 2003).

G. Camarillo & M.-A. Garcia-Martin, *The 3G IP Multimedia Subsystem (IMS): Merging the Internet and the Cellular Worlds*, 2nd edition (Wiley, 2005).

C. Chevallier, C. Brunner, A. Garavaglia, K. Murray & K. Baker, *WCDMA (UMTS) Deployment Handbook: Planning and Optimization Aspects* (Wiley, 2006).

E. Dahlman, S. Parkvall, J. Sköld & P. Beming, *3G Evolution: HSPA and LTE for Mobile Broadband* (Academic Press, 2007).

C. Johnson, *Radio Access Networks for UMTS: Principles and Practice* (Wiley, 2008).

J. Korhonen, *Introduction to 3G Mobile Communications*, 2nd edition (Artech, 2003).

P. Lescuyer, translated by F. Bott, *UMTS: Origins, Architecture and the Standard* (Springer-Verlag, 2004).

A. Richardson, *WCDMA Design Handbook* (Cambridge University Press, 2005).

R. Tanner & J. Woodard, *WCDMA Requirements and Practical Design* (Wiley, 2004).

There are a number of websites containing useful information about UMTS. The website of the 3rd Generation Partnership Project is www.3gpp.org. This site contains all the specifications for GSM and UMTS, together with a large amount of background material about the project. Because there are several hundred specifications, it can sometimes be hard for people to find the one they require, so the author maintains a visual interface to the UMTS specifications at www.chriscoxcommunications.co.uk. Two excellent sources of technical information and market data are the websites of the UMTS Forum, www.umts-forum.org, and 3G Americas, www.3gamericas.org.

# Abbreviations

| | |
|---|---|
| 1G | first generation |
| 1×EV-DO | 1×Evolution Data Optimised |
| 1×RTT | 1×Radio Transmission Technology |
| 2G | second generation |
| 3G | third generation |
| 3GPP | 3rd Generation Partnership Project |
| 3GPP2 | 3rd Generation Partnership Project 2 |
| 4G | fourth generation |
| A/D | analogue/digital |
| AAL | ATM adaptation layer |
| ABMF | account balance management function |
| AICH | acquisition indicator channel |
| ALCAP | access link control application protocol |
| AM | acknowledged mode |
| AMPS | Advanced Mobile Phone System |
| AMR | adaptive multi rate |
| API | application programming interface |
| APN | access point name |
| ARIB | Association of Radio Industries and Businesses |
| ARQ | automatic repeat request |
| AS | access stratum |
| ASIC | application specific integrated circuit |
| AT | attention |
| ATIS | Alliance for Telecommunications Industry Solutions |
| ATM | asynchronous transfer mode |
| AuC | authentication centre |
| BCCH | broadcast control channel |
| BCH | broadcast channel |

| | |
|---|---|
| BER | bit error ratio |
| BGCF | border gateway control function |
| BICC | bearer independent call control |
| BLER | block error ratio |
| BMC | broadcast/multicast control |
| BM-SC | broadcast/multicast service centre |
| BPSK | binary phase shift keying |
| BSC | base station controller |
| BSS | base station subsystem |
| BSSAP | base station subsystem application part |
| BSSAP+ | base station subsystem application part plus |
| BTS | base transceiver station |
| CAMEL | customised application for mobile network enhanced logic |
| CAP | CAMEL application part |
| CBS | cell broadcast service |
| CC | call control |
| CC/PP | composite capability/preference profiles |
| CCCH | common control channel |
| CCSA | China Communications Standards Association |
| CDF | charging data function |
| CDMA | code division multiple access |
| CDR | charging data record |
| CFN | connection frame number |
| CGF | charging gateway function |
| CM | connection management |
| CN | core network |
| CPICH | common pilot channel |
| CQI | channel quality indicator |
| CRC | cyclic redundancy check |
| CRNC | controlling radio network controller |
| C-RNTI | cell radio network temporary identity |
| CS | circuit switched |
| CSCF | call session control function |
| CS-MGW | circuit switched media gateway |

| | |
|---|---|
| CTCH | common traffic channel |
| CTF | charging trigger function |
| | |
| D/A | digital/analogue |
| D-AMPS | Digital AMPS |
| dB | decibels |
| dBm | decibels relative to 1 milliwatt |
| DCCH | dedicated control channel |
| DCH | dedicated channel |
| DiffServ | differentiated services |
| DL | downlink |
| DPCCH | dedicated physical control channel |
| DPDCH | dedicated physical data channel |
| DRNC | drift radio network controller |
| DRX | discontinuous reception |
| DTCH | dedicated traffic channel |
| | |
| E-AGCH | E-DCH absolute grant channel |
| E-DCH | enhanced dedicated channel |
| EDGE | Enhanced Data Rates for Global Evolution |
| E-DPCCH | E-DCH dedicated physical control channel |
| E-DPDCH | E-DCH dedicated physical data channel |
| E-HICH | E-DCH hybrid ARQ indicator channel |
| EIR | equipment identity register |
| eNodeB | evolved Node B |
| EPC | evolved packet core |
| E-RGCH | E-DCH relative grant channel |
| ETSI | European Telecommunications Standards Institute |
| E-UTRAN | evolved UMTS radio access network |
| | |
| FACH | forward access channel |
| FBI | feedback information |
| FCC | Federal Communications Commission |
| FDD | frequency division duplex |
| FDMA | frequency division multiple access |
| FFT | fast Fourier transform |

| | |
|---|---|
| GERAN | GSM EDGE radio access network |
| GGSN | gateway GPRS support node |
| GMM | GPRS mobility management |
| GMSC | gateway mobile switching centre |
| GPRS | general packet radio service |
| GPS | Global Positioning System |
| GRX | GPRS roaming exchange |
| GSM | Global System for Mobile Communications |
| gsmSCF | GSM service control function |
| gsmSSF | GSM service switching function |
| GTP | GPRS tunnelling protocol |
| GTP-C | GPRS tunnelling protocol control part |
| GTP-U | GPRS tunnelling protocol user part |
| GUP | generic user profile |
| HARQ | hybrid automatic repeat request |
| HCS | hierarchical cell structure |
| HLR | home location register |
| HSDPA | high speed downlink packet access |
| HS-DPCCH | high speed dedicated physical control channel |
| HS-DSCH | high speed downlink shared channel |
| HSPA | high speed packet access |
| HSPA+ | high speed packet access evolution |
| HS-PDSCH | high speed physical downlink shared channel |
| HSS | home subscriber server |
| HS-SCCH | high speed shared control channel |
| HSUPA | high speed uplink packet access |
| HTTP | hypertext transfer protocol |
| I | in-phase |
| I-CSCF | interrogating call session control function |
| iDEN | Integrated Digital Enhanced Network |
| IEEE | Institute of Electrical and Electronics Engineers |
| IETF | Internet Engineering Task Force |
| IMEI | international mobile equipment identity |
| IM-MGW | IMS media gateway |
| IMS | IP multimedia subsystem |

| | |
|---|---|
| IMSI | international mobile subscriber identity |
| IMT | International Mobile Telecommunications |
| IP | Internet protocol |
| IPDL | idle period in the downlink |
| ISDN | Integrated Services Digital Network |
| ISI | inter symbol interference |
| ISIM | IP multimedia services identity module |
| ISUP | ISDN user part |
| ITU | International Telecommunication Union |
| JPEG | Joint Photographic Experts Group |
| LA | location area |
| LAI | location area identity |
| LAN | local area network |
| LCS | location services |
| LTE | Long Term Evolution |
| MAC | medium access control |
| MAP | mobile application part |
| MBMS | multimedia broadcast/multicast service |
| MCC | mobile country code |
| Mcps | million chips per second |
| ME | mobile equipment |
| MEGACO | media gateway control protocol |
| MExE | mobile execution environment |
| MGCF | media gateway control function |
| MGW | media gateway |
| MIB | master information block |
| MIMO | multiple input multiple output |
| MM | mobility management |
| MMS | multimedia messaging service |
| MNC | mobile network code |
| MPEG | Moving Pictures Experts Group |
| MRFC | media resource function controller |
| MRFP | media resource function processor |
| MS | mobile station |
| MSC | mobile switching centre |

| | |
|---|---|
| MS-ISDN | mobile station ISDN number |
| MT | mobile termination |
| MTP | message transfer part |
| | |
| NAS | non-access stratum |
| NBAP | Node B application part |
| NMT | Nordic Mobile Telephone |
| | |
| OCF | online charging function |
| OCS | online charging system |
| OFDMA | orthogonal frequency division multiple access |
| OMA | Open Mobile Alliance |
| OSA | open service access |
| OSI | open systems interconnection |
| OTDOA | observed time difference of arrival |
| | |
| PCCH | paging control channel |
| PCCPCH | primary common control physical channel |
| PCG | project co-ordination group |
| PCH | paging channel |
| PCM | pulse code modulation |
| P-CSCF | proxy call session control function |
| PDC | Personal Digital Cellular |
| PDCP | packet data convergence protocol |
| PDG | packet data gateway |
| PDN | packet data network/public data network |
| PDP | packet data protocol |
| PDU | protocol data unit |
| PICH | paging indicator channel |
| PIN | personal identification number |
| PLMN | public land mobile network |
| PLMN-ID | public land mobile network identity |
| PoC | push-to-talk over cellular |
| PPP | point-to-point protocol |
| PRACH | physical random access channel |
| PS | packet switched |
| P-SCH | primary synchronisation channel |
| PSS | packet switched streaming |

| | |
|---|---|
| PSTN | public switched telephone network |
| P-TMSI | packet temporary mobile subscriber identity |
| | |
| Q | quadrature |
| QAM | quadrature amplitude modulation |
| QoS | quality of service |
| QPSK | quadrature phase shift keying |
| | |
| RA | routing area |
| RAB | radio access bearer |
| RACH | random access channel |
| RAI | routing area identity |
| RANAP | radio access network application part |
| RB | radio bearer |
| RF | rating function |
| RFCI | RAB sub-flow combination indicator |
| RLC | radio link control |
| RLP | radio link protocol |
| RNC | radio network controller |
| RNS | radio network subsystem |
| RNSAP | radio network subsystem application part |
| RNTI | radio network temporary identity |
| RRC | radio resource control |
| RTP | real time protocol |
| | |
| SAE | System Architecture Evolution |
| SC | service centre |
| SCCP | signalling connection control part |
| SCCPCH | secondary common control physical channel |
| SC-FDMA | single carrier frequency division multiple access |
| SCH | synchronisation channel |
| SCS | service capability server |
| S-CSCF | serving call session control function |
| SDP | session description protocol |
| SDU | service data unit |
| SFN | system frame number |
| SGSN | serving GPRS support node |
| SIB | system information block |

| | |
|---|---|
| SID | silence information descriptor |
| SIM | subscriber identity module |
| SIP | session initiation protocol |
| SIR | signal-to-interference ratio |
| SLF | subscriber location function |
| SM | session management |
| SMS | short message service |
| SMS-GMSC | SMS gateway MSC |
| SMS-IWMSC | SMS interworking MSC |
| SMTP | simple mail transfer protocol |
| SNR | signal-to-noise ratio |
| SRB | signalling radio bearer |
| SRNC | serving radio network controller |
| SS | supplementary services |
| SS7 | signalling system number 7 |
| S-SCH | secondary synchronisation channel |
| TACS | Total Access Communication System |
| TAF | terminal adaptation function |
| TAP | transferred account procedure |
| TCAP | transaction capabilities application part |
| TCP | transmission control protocol |
| TDD | time division duplex |
| TDMA | time division multiple access |
| TD-SCDMA | time division synchronous code division multiple access |
| TE | terminal equipment |
| TF | transport format |
| TFC | transport format combination |
| TFCI | transport format combination indicator |
| TFCS | transport format combination set |
| TFI | transport format indicator |
| TFS | transport format set |
| TFT | traffic flow template |
| TG | transmission gap |
| TM | transparent mode |
| TMSI | temporary mobile subscriber identity |

| | |
|---|---|
| TPC | transmit power control |
| TR | technical report |
| TS | technical specification |
| TSG | technical specification group |
| TTA | Telecommunications Technology Association |
| TTC | Telecommunication Technology Committee |
| TTI | transmission time interval |
| TUP | telephone user part |
| UDP | user datagram protocol |
| UE | user equipment |
| UICC | universal integrated circuit card |
| UL | uplink |
| UM | unacknowledged mode |
| UMB | Ultra Mobile Broadband |
| UMTS | Universal Mobile Telecommunication System |
| URA | UTRAN registration area |
| U-RNTI | UTRAN radio network temporary identity |
| USAT | USIM application toolkit |
| USB | universal serial bus |
| USIM | universal subscriber identity module |
| UTRAN | UMTS terrestrial radio access network |
| VHE | virtual home environment |
| VLR | visitor location register |
| VoIP | voice over IP |
| WAG | WLAN access gateway |
| WAP | wireless application protocol |
| W-CDMA | wideband code division multiple access |
| WiMAX | Worldwide Interoperability for Microwave Access |
| WLAN | wireless local area network |

# Index

Note: Illustrations are indicated by page numbers in italics.